DATA STRUCTURES
AND
PROGRAM DESIGN
USING PYTHON

DATA STRUCTURES AND PROGRAM DESIGN USING PYTHON

A Self-Teaching Introduction

Dheeraj Malhotra, PhD
Neha Malhotra, PhD

MERCURY LEARNING AND INFORMATION
Dulles, Virginia
Boston, Massachusetts
New Delhi

Publisher: David Pallai
MERCURY LEARNING AND INFORMATION
22841 Quicksilver Drive
Dulles, VA 20166
info@merclearning.com
www.merclearning.com
(800) 232-0223

D. Malhotra and N. Malhotra. *Data Structures and Program Design Using Python.*
ISBN: 978-1-68392-639-9

The publisher recognizes and respects all marks used by companies, manufacturers, and developers as a means to distinguish their products. All brand names and product names mentioned in this book are trademarks or service marks of their respective companies. Any omission or misuse (of any kind) of service marks or trademarks, etc. is not an attempt to infringe on the property of others.

Library of Congress Control Number: 2020946121

2021223321 Printed on acid-free paper in the United States of America.

Our titles are available for adoption, license, or bulk purchase by institutions, corporations, etc. For additional information, please contact the Customer Service Dept. at (800) 232-0223(toll free). Digital versions of our titles are available at: *www.academiccourseware.com* and other electronic vendors.

*Dedicated to our
loving parents and beloved students*

CONTENTS

PREFACE

Data structures are the building blocks of computer science. The objective of this text is to emphasize the fundamentals of data structures as an introductory subject. It is designed for beginners who would like to learn the basics of data structures and their implementation using the Python programming language. With this focus in mind, we present various fundamentals of the subject, well supported with real-world analogies to enable a quick understanding of the technical concepts and to help the reader in quickly identifying appropriate data structures to solve specific, practical problems. This book will serve the purpose of a text or reference book and will be of immense help especially to undergraduate or graduate students of various courses in information technology, engineering, computer applications, and information sciences.

Key Features:

- *Practical Applications*: Real-world analogies as practical applications are given throughout the text to quickly understand and connect the fundamentals of data structures with day to day, real-world scenarios. This approach, in turn, will assist the reader in developing the capability to identify the most appropriate and efficient data structure for solving a specific, real-world problem.

- *Frequently Asked Questions*: Frequently asked theoretical or practical questions are integrated throughout the content of the book, within related topics to assist readers in grasping the subject.

- *Algorithms and Programs*: To better understand the fundamentals of data structures at a generic level-followed by their implementation in Python, syntax independent algorithms, as well as implemented programs in Python, are discussed throughout the book. This presentation will assist the reader in getting both algorithms and their corresponding implementation within a single book.

- *Numerical and Conceptual Exercises*: To assist the reader in developing a strong foundation of the subject, various numerical and conceptual problems are included throughout the text.
- *Multiple Choice Questions*: To assist students for placement-oriented exams in various IT fields, several exercises are suitably chosen and are given in an MCQ format.

Dr. Dheeraj Malhotra
Dr. Neha Malhotra
September 2020

ACKNOWLEDGMENTS

We are indeed grateful to Chairman - Dr. S.C. Vats, Vice Chairman - Mr. Suneet Vats, Chairperson VSIT - Prof. Sidharth Mishra, and Dean VSIT- Prof. Supriya Madan of our employer institute - Vivekananda Institute of Professional Studies (GGS IP University). They are always a source of inspiration for us, and we feel honored because of their faith in us.

We also take this opportunity to extend our gratitude to our mentors Prof. O.P. Rishi (University of Kota), Dr. Sushil Chandra (DRDO, GOI), Prof. Udyan Ghose (GGS IP University), and Prof. M.N. Hoda (BVICAM) for their motivation to execute this project.

We are profoundly thankful to Mr. Sahil Pathak (VIPS, GGSIPU) and Mr. Deepanshu Gupta (Tech Mahindra Ltd.) for helping us in proofreading and compiling the codes in this manuscript.

It is not possible to complete a book without the support of a publisher. We are thankful to David Pallai and Jennifer Blaney of Mercury Learning and Information for their enthusiastic involvement throughout the tenure of this project.

Our heartfelt regards to our parents, siblings and family members who cheered us in good times and encouraged us in bad times.

Lastly, we have always felt inspired by our readers, especially in the USA, Canada, and India. Their utmost love and positive feedback for our first three titles of *Data Structures using C*, …*C++*, and …*Java*, all published with MLI, helped us to improve the current title further.

Dr. Dheeraj Malhotra
Dr. Neha Malhotra
September 2020

## CHAPTER 1

INTRODUCTION TO DATA STRUCTURES

1.1 INTRODUCTION

A data structure is an efficient way of storing and organizing the data elements in a computer's memory. *Data* means a value or a collection of values. *Structure* refers to a method of organizing the data. The mathematical or logical representation of data in the memory is referred to as a *data structure*. The objective of a data structure is to store, retrieve, and update the data efficiently. A data structure can be considered as all the elements grouped under one name. The data elements are called *members*, and they can be of different types. Data structures are used in almost every program and software system. There are various kinds of data structures that are suited for different types of applications. Data structures are the building blocks of a program. For a program to run efficiently, a programmer must choose the appropriate data structures. A data structure is a crucial part of data management. As the name suggests, *data management* is a task that includes different activities, like the collection of data and the organization of data into structures. Data structures are used in stacks, queues, arrays, binary trees, linked lists, and hash tables.

A data structure helps us to understand the relationship of one element to another element and organize it within the memory. It is a mathematical or logical representation or organization of data in the memory. Data structures are extensively applied in the following areas:

- Compiler Design
- Database Management Systems (DBMS)

- Artificial Intelligence
- Network and Numerical Analysis
- Statistical Analysis Packages
- Graphics
- Operating Systems (OS)
- Simulations

There are many applications in which different data structures are used for their operations. Some data structures sacrifice speed for the efficient utilization of memory, while others sacrifice memory utilization and result in a faster speed. In today's world, programmers aim not just to build a program, but to build an effective program. As previously discussed, for a program to be efficient, a programmer must choose the appropriate data structures. Hence, data structures are classified into various types. Now, let us discuss and learn about different types of data structures.

Frequently Asked Questions

1. Define the term "data structure."

Answer:

A data structure is an organization of data in a computer's memory or disk storage. In other words, a logical or mathematical model of a particular organization of data is called a data structure. A data structure in computer science is also a way of storing data in a computer so that it can be used efficiently. An appropriate data structure allows a variety of important operations to be performed using both resources, that is, the memory space and execution time, efficiently.

1.2 TYPES OF DATA STRUCTURES

Data structures are classified into various types.

1.2.1 Linear and Non-Linear Data Structures

A *linear data structure* is one in which the data elements are stored in a linear, or sequential, order; that is, data is stored in consecutive memory locations.

A linear data structure can be represented in two ways; either it is represented by a linear relationship between various elements utilizing consecutive memory locations as in the case of arrays, or it may be represented by a linear relationship between the elements utilizing links from one element to another as in the case of linked lists. Examples of linear data structures include arrays, linked lists, stacks, and queues.

A *non-linear data structure* is one in which the data is not stored in any sequential order or consecutive memory locations. The data elements in this structure are represented by a hierarchical order. Examples of non-linear data structures include graphs and trees.

1.2.2 Static and Dynamic Data Structures

A *static data structure* is a collection of data in memory that is fixed in size and cannot be changed during runtime. The memory size must be known in advance, as the memory cannot be reallocated later in a program. One example is an *array*.

A *dynamic data structure* is a collection of data in which memory can be reallocated during the execution of a program. The programmer can add or remove elements according to his/her need. Examples include linked lists, graphs, and trees.

1.2.3 Homogeneous and Non-Homogeneous Data Structures

A *homogeneous data structure* is one that contains data elements of the same type (for example, arrays).

A *non-homogeneous data structure* contains data elements of different types (for example, structures).

1.2.4 Primitive and Non-Primitive Data Structures

Primitive data structures are the fundamental data structures or predefined data structures that are supported by a programming language. Examples of primitive data structure types are integer, float, and char.

Non-primitive data structures are comparatively more complicated data structures that are created using primitive data structures. Examples of non-primitive data structures are arrays, files, linked lists, stacks, and queues.

The classification of different data structures is shown in Figure 1.1.

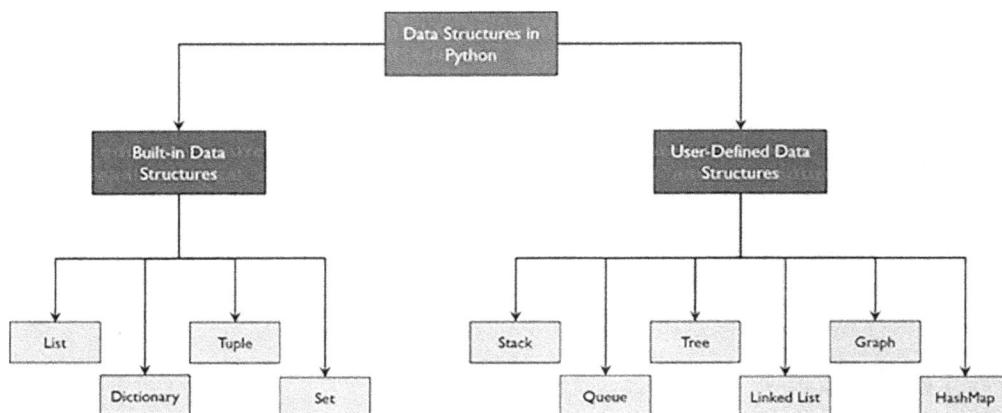

FIGURE 1.1 Classification of different data structures

Python supports various data structures. We now introduce all these data structures, and they are discussed in detail in the upcoming chapters.

Frequently Asked Questions

2. What is the difference between primitive data structures and non-primitive data structures?

Answer:

The data structures that are typically directly operated upon by machine-level instructions, that is, the fundamental data types such as int, float, and char, are known as primitive data structures. The data structures that are not fundamental are called non-primitive data structures.

Frequently Asked Questions

3. What is the difference between linear and non-linear data structures?

Answer:

The main difference between linear and non-linear data structures lies in the way in which data elements are organized. In the linear data structure, elements are organized sequentially, and therefore they are easy to implement in a computer's memory. In non-linear data structures, a data element can be attached to several other data elements to represent specific relationships existing among them.

1.2.5 Arrays/Lists

The array structure looks very similar to Python's list structure. That's because the two structures are both sequences that are composed of multiple sequential elements that can be accessed by position. But there are two major differences between the array and the list. First, an array has a limited number of operations, which commonly include those for array creation, reading a value from a specific element, and writing a value to a specific element. The list provides a large number of operations for working with the content of the list. Second, the list can grow and shrink during execution as elements are added or removed while the size of an array cannot be changed after it has been created.

Python's list structure is a mutable sequence container that can change size as items are added or removed. It is an abstract data type that is implemented using an array structure to store the items contained in the list.

In Python, a list is declared using the following syntax:

Syntax – pyList = [4, 12, 2, 34, 17]

1.2.6 Stacks

A *stack* is a collection of objects that are inserted and removed according to the Last-In, First-Out (LIFO) principle. A user may insert objects into a stack at any time, but may only access or remove the most recently inserted object that remains (at the so-called "top" of the stack). The name "stack" is derived from the metaphor of a stack of plates in a spring-loaded, cafeteria plate dispenser. In this case, the fundamental operations involve the "pushing" and "popping" of plates on the stack. When we need a new plate from the dispenser, we "pop" the top plate off the stack, and when we add a plate, we "push" it down on the stack to become the new top plate.

> **Practical Application:**
>
> A real-life example of a stack is a pile of plates arranged on a table. A person will pick up the first plate from the top of the stack.

The Stack ADT can be implemented in several ways. The two most common approaches to implement Stack ADT in Python include the use of a Python list and a linked list. The choice depends on the type of application involved.

1.2.7 Queues

Another fundamental data structure is the *queue*. It is a close cousin of the stack, as a queue is a collection of objects that are inserted and removed according to the First-In, First-Out (FIFO) principle. That is, elements can be inserted at any time, but only the element that has been in the queue the longest can be next removed. We usually say that elements enter a queue at the back and are removed from the front. A metaphor for this terminology is a line of people waiting to get on an amusement park ride. People waiting for such a ride enter at the back of the line and get on the ride from the front of the line.

Practical Application:

For a simple illustration of a queue, imagine there is a line of people standing at the bus stop and waiting for the bus. The first person standing in the line will get into the bus first.

The Queue ADT can be implemented in several ways. The two most common approaches in Python include the use of a Python list and a linked list. The choice depends on the type of application involved.

1.2.8 Linked Lists

The major drawback of the array is that the size or the number of elements must be known in advance. Thus, this drawback gave rise to the new concept of a linked list. A *linked list* is a linear collection of data elements. These data elements are called *nodes*, which store the address of the next node. A linked list is a sequence of nodes in which each node contains one or more than one data field and an address field that stores the address of the next node. Linked lists are dynamic; that is, memory is allocated when required.

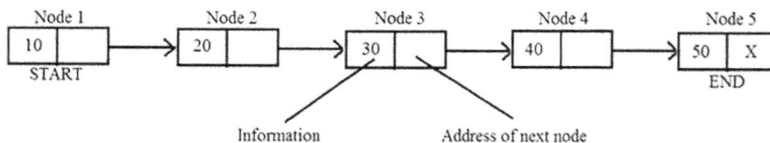

FIGURE 1.2 Memory representation of a linked list

Figure 1.2 shows a linked list in which each node is divided into two slots:

1. The first slot contains information/data.

2. The second slot contains the address of the next node.

Practical Application:

A simple real-life example is a train; here, each train car is connected to the previous one and next one (except the first car (the engine) and the last car (the coach)).

The address part of the last node stores a special value called NULL, which denotes the end of the linked list. The advantage of a linked list over arrays is that now it is easier to insert and delete data elements, as we don't have to do shifting each time. Yet searching for an element is difficult. More time is required to search for an element, and it requires a large amount of memory space. Hence, linked lists are used where a collection of data elements is required but the number of data elements in the collection is not known to us in advance.

Frequently Asked Questions

4. Define the term "linked list."

Answer:

A linked list or one-way list is a linear collection of data elements called nodes, which give a linear order. It is a popular dynamic data structure. The nodes in the linked list are not stored in consecutive memory locations. For every data item in a node of the linked list, there is an associated address field that gives the address location of the next node in the linked list.

1.2.9 Trees

A *tree* is a popular non-linear data structure in which the data elements or the nodes are represented in a hierarchical order. Here, one of the nodes is shown as the root node of the tree, and the remaining nodes are partitioned into two disjointed sets such that each set is a part of a sub-tree. A tree makes the search process very easy, and its recursive programming makes a program optimized and easy to understand.

A binary tree is the simplest form of a tree. A *binary tree* consists of a root node and two sub-trees known as the left sub-tree and the right sub-tree, where both sub-trees are also binary trees. Each node in a tree consists of three parts, that is, the extreme left part stores the address of the left sub-tree, the middle part consists of the data element, and the extreme right part stores the address of the right sub-tree. The root is the topmost element of the tree. When there are no nodes in a tree, that is, when ROOT = NULL, then it is called an *empty tree*.

For example, consider a binary tree where R is the root node of the tree. LEFT and RIGHT are the left and right sub-trees of R, respectively. Node A is designated as the root node of the tree. Nodes B and C are the left and right child of A, respectively. Nodes B, D, E, and G constitute the left sub-tree of the root. Similarly, nodes C, F, H, and I constitute the right sub-tree of the root.

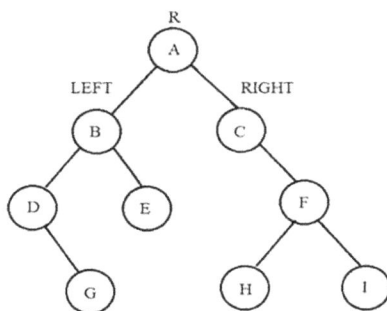

FIGURE 1.3 A binary tree

Advantages of a tree

1. The searching process is very fast in trees.

2. The insertion and deletion of the elements is easier compared to other data structures.

Frequently Asked Questions

5. Define the term "binary tree."

Answer:

A binary tree is a hierarchal data structure in which each node has at most two children, that is, the left and right child. In a binary tree, the degree of each node can be at most two. Binary trees are used to implement binary search trees, which are used for efficient searching and sorting. A variation of BST is an AVL tree, where the height of the left and right subtree differs by one. A binary tree is a popular subtype of a k-ary tree, where k is 2.

1.2.10 Graphs

A *graph* is a general tree with no parent-child relationship. It is a non-linear data structure that consists of vertices, also called nodes, and the edges that connect those vertices. In a graph, any complex relationship can exist. A graph G may be defined as a finite set of V vertices and E edges. Therefore, G = (V, E) where V is the set of vertices and E is the set of edges. Graphs are used in various applications of mathematics and computer science. Unlike a root node in trees, graphs don't have root nodes; rather, the nodes can be connected to any node in the graph. Two nodes are called *neighbors* when they are connected via an edge.

> **Practical Application:**
>
> A real-life example of a graph can be seen in workstations where several computers are joined to one another via network connections.

For example, consider a graph G with six vertices and eight edges. Here, Q and Z are neighbors of P. Similarly, R and T are neighbors of S.

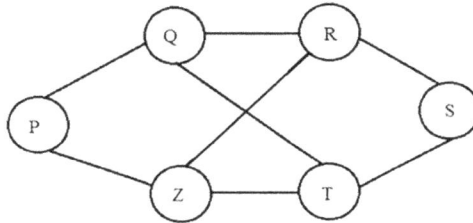

FIGURE 1.4 A graph

1.3 OPERATIONS ON DATA STRUCTURES

Here, we discuss various operations that are performed on data structures.

- **Creation** – This is the process of creating a data structure. The declaration and initialization of the data structure are done here. It is the first operation.

- **Insertion** – This is the process of adding new data elements in the data structure, for example, to add the details of an employee who has recently joined an organization.

- **Deletion** – This is the process of removing a particular data element from the given collection of data elements, for example, to remove the name of an employee who has left the company.

- **Updating** – This is the process of modifying the data elements of a data structure. For example, if the address of a student is changed, it should be updated.

- **Searching** – This is used to find the location of a particular data element or all the data elements with the help of a given key, for example, to find the names of people who live in New York.

- **Sorting** – This is the process of arranging the data elements in some order, that is, either ascending or descending order. An example is arranging the names of students of a class in alphabetical order.

- **Merging** – This is the process of combining the data elements of two different lists to form a single list of data elements.
- **Traversal** – This is the process of accessing each data element exactly once so that it can be processed. An example is to print the names of all the students of a class.
- **Destruction** – This is the process of deleting the entire data structure. It is the last operation in the data structure.

1.4 ALGORITHMS

An *algorithm* is a systematic set of instructions combined to solve a complex problem. It is a step-by-finite-step sequence of instructions, each of which has a clear meaning and can be executed in a minimum amount of effort in finite time. In general, an algorithm is a blueprint for writing a program to solve the problem. Once we have a blueprint of the solution, we can easily implement it in any high-level language like C, C++, or Python. It divides the problem into a finite number of steps. An algorithm written in a programming language is known as a *program*. A computer is a machine with no brain or intelligence. Therefore, the computer must be instructed to perform a given task in unambiguous steps. Hence, a programmer must define his problem in the form of an algorithm written in English. Thus, such an algorithm should have the following features:

1. An algorithm should be simple and concise.
2. It should be efficient and effective.
3. It should be free of ambiguity; that is, the logic must be clear.

    Similarly, an algorithm must have the following characteristics:

- **Input** – It reads the data of the given problem.
- **Output** – The desired result must be produced.
- **Process/Definiteness** – Each step or instruction must be unambiguous.
- **Effectiveness** – Each step should be accurate and concise. The desired result should be produced within a finite time.
- **Finiteness** – The number of steps should be finite.

1.4.1 Developing an Algorithm

To develop an algorithm, some steps are necessary:

1. Defining or understanding the problem.

2. Identifying the result or output of the problem.

3. Identifying the inputs required by the problem and choosing the best input.

4. Designing the logic from the given inputs to get the desired output.

5. Testing the algorithm for different inputs.

6. Repeating the previous steps until it produces the desired result for all the inputs.

1.5   APPROACHES FOR DESIGNING AN ALGORITHM

A complicated algorithm is divided into smaller units called *modules*. These modules are further divided into sub-modules. Thus, in this way, a complex algorithm can easily be solved. The process of dividing an algorithm into modules is called *modularization*. There are two popular approaches for designing an algorithm:

• Top-Down Approach
• Bottom-Up Approach

Now let us understand both approaches.

1. **Top-Down Approach**–A *top-down approach* states that the complex/complicated problem/algorithm should be divided into a smaller number of one or more modules. These smaller modules are further divided into one or more sub-modules. This process of decomposition is repeated until we achieve the desired output of module complexity. A top-down approach starts from the topmost module, and the modules are incremented accordingly until a level is reached where we don't require any more sub-modules, that is, the desired level of complexity is achieved.

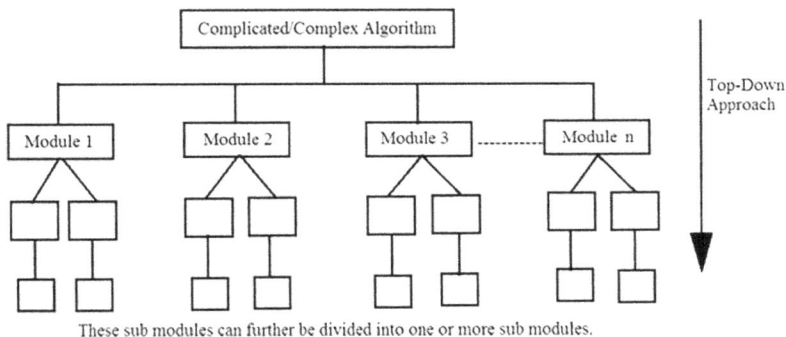

These sub modules can further be divided into one or more sub modules.

FIGURE 1.5 Top-down approach

2. **Bottom-Up Approach**–A bottom-up algorithm design approach is the opposite of a top-down approach. In this kind of approach, we first start with designing the basic modules and proceed further toward designing the high-level modules. The sub-modules are grouped to form a module of a higher level. Similarly, all high-level modules are grouped to form more high-level modules. Thus, this process of combining the sub-modules is repeated until we obtain the desired output of the algorithm.

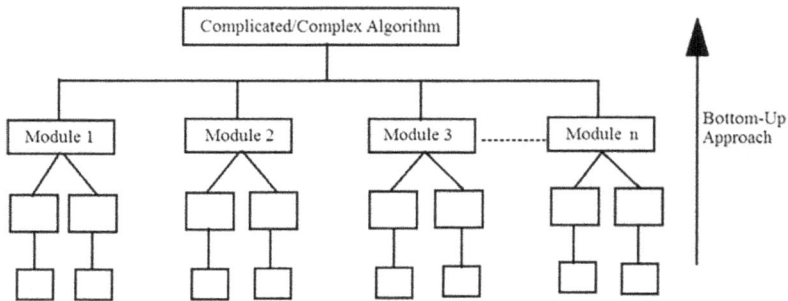

FIGURE 1.6 Bottom-up approach

1.6 ANALYZING AN ALGORITHM

An algorithm can be analyzed by two factors: space and time. We should develop an algorithm that makes the best use of both these resources. Analyzing an algorithm measures the efficiency of the algorithm. The efficiency of the algorithm is measured in terms of the speed and time complexity. The complexity of an algorithm is a function that measures the space and time used by an algorithm in terms of input size.

Time Complexity–The *time complexity* of an algorithm is the amount of time taken by an algorithm to run the program completely. It is the runtime of the program. The time complexity of an algorithm depends upon the input size. The time complexity is commonly represented by using big O notation. For example, the time complexity of a linear search is $O(n)$.

Space Complexity–The *space complexity* of an algorithm is the amount of memory space required to run the program completely. The space complexity of an algorithm depends upon the input size.

Time Complexity is categorized into three types:

1. **Best Case Running Time**–The performance of the algorithm is best under optimal conditions. For example, the best case for a binary search occurs when the desired element is the middle element of the list. Another

example is that of sorting; that is, if the elements are already sorted in a list, then the algorithm will execute in the best time.

2. **Average Case Running Time**–This denotes the behavior of an algorithm when the input is randomly drawn from a given collection or distribution. It is an estimate of the running time for "average" input. It is usually assumed that all inputs of a given size are likely to occur with equal probability.

3. **Worst Case Running Time**–The behavior of the algorithm during the worst possible case of the input instance. The worst case running time of an algorithm is an upper bound on the running time for any input. For example, the worst case for a linear search occurs when the desired element is the last element in the list or the element does not exist in the list.

Frequently Asked Questions

6. Define time complexity.

Answer:

Time complexity is a measure that evaluates the count of the operations performed by a given algorithm as a function of the size of the input. It is the approximation of the number of steps necessary to execute an algorithm. It is commonly represented with asymptotic notation, that is, the O(g) notation, also known as big O notation, where g is the function of the size of the input data.

1.6.1 Time-Space Trade-Off

In computer science, the *time-space trade-off* is a way of solving a particular problem either in less time and more memory space or in more time and less memory space. But if we talk in practical terms, designing such an algorithm in which we can save both space and time is a challenging task. So, we can use more than one algorithm to solve a problem. One may require less time, and the other may require less memory space to execute. Therefore, we sacrifice one thing for the other. Hence, there exists a time-space or time-memory trade-off between algorithms. The time-space trade-off gives the programmer a rational choice from an informed point of view. If time is a big concern for a programmer, then she might choose a program that takes less or the minimum time to execute. If space is a prime concern for a programmer, then, she might choose a program that takes less memory space to execute at the cost of more time.

1.7 ABSTRACT DATA TYPES

An *Abstract Data Type (ADT)* is a popular mathematical model of data objects that defines a data type along with various functions that operate on these objects. To understand the meaning of an abstract data type, we can break the term into two parts, that is, "data type" and "abstract." The data type of a variable is a collection of values that a variable can take. There are various data types in Python (such as integer, float, character, long, and double). When we talk about the term "abstract" in the context of data structures, it means "apart from detailed specifications." It can be considered as a description of the data in a structure with a list of operations to be executed on the data within the structure. Thus, an abstract data type is the specification of a data type that specifies the mathematical and logical model of the data type. For example, when we use stacks and queues, our prime concern is only with the data type and the operations to be performed on those structures. We are not worried about how the data will be stored in the memory. Also, we don't bother about how the push () and pop () operations work. We just know that we have two functions available to us, so we have to use them for insertion and deletion operations.

1.8 BIG O NOTATION

The performance of an algorithm, that is, time and space requirements, can be easily compared with other competitive algorithms using asymptotic notations such as the big O notation, the Omega notation, and the Theta notation. The algorithmic complexity can be easily approximated using asymptotic notations by simply ignoring the implementation-dependent factors. For instance, we can compare various available sorting algorithms using the big O notation or any other asymptotic notation.

Big O notation is a popular analysis characterization scheme because it provides an upper bound on the complexity of an algorithm. In big O, $O(g)$ is representative of the class of all functions that grow no faster than g. Therefore, if $f(n) = O(g(n))$, then $f(n) <= c(g(n))$ for all $n > n_0$, where n_0 represents a threshold and c represents a constant.

An algorithm with $O(1)$ complexity is referred to as a constant computing time algorithm. Similarly, an algorithm with $O(n)$ complexity is referred to as a linear algorithm, an algorithm with $O(n^2)$ complexity is referred to as a quadratic algorithm, an algorithm with $O(2^n)$ complexity is referred to as an exponential time algorithm, an algorithm with $O(n^k)$ complexity is referred to as a polynomial-time algorithm, and an algorithm with $O(\log n)$ complexity is referred to as a logarithmic time algorithm.

An algorithm with the complexity of the order of $O(\log_2 n)$ is considered as one of the best algorithms, while an algorithm with the complexity of the order of $O(2^n)$ is considered as the worst algorithm. The complexity of computations or the number of iterations required in various types of functions may be compared as follows:

$$O(\log_2 n) < O(n) < O(n \log_2 n) < O(n^2) < O(n^3) < O(2^n)$$

1.9 SUMMARY

- A data structure determines a way of storing and organizing the data elements in a computer's memory. Data means a value or a collection of values. Structure refers to a way of organizing the data. The mathematical or logical representation of data in the memory is referred to as a data structure.

- Data structures are classified into various types, which include linear and non-linear data structures, primitive and non-primitive data structures, static and dynamic data structures, and homogeneous and non-homogeneous data structures.

- A linear data structure is one in which the data elements are stored in a linear or sequential order; that is, data is stored in consecutive memory locations. A non-linear data structure is one in which the data is not stored in any sequential order or consecutive memory locations.

- A static data structure is a collection of data in memory that is fixed in size and cannot be changed during runtime. A dynamic data structure is a collection of data in which memory can be reallocated during the execution of a program.

- Primitive data structures are fundamental data structures or predefined data structures that are supported by a programming language. Non-primitive data structures are comparatively more complicated data structures that are created using primitive data structures.

- A homogeneous data structure is one that contains all data elements of the same type. A non-homogeneous data structure contains data elements of different types.

- Python's list structure is a mutable sequence container that can change size as items are added or removed. It is an abstract data type that is implemented using an array structure to store the items contained in the list.

- A queue is a linear collection of data elements in which the element inserted first will be the element taken out first, that is, it is a FIFO data structure. A queue is a linear data structure in which the first element is inserted from one end, called the REAR end, and the deletion of the element takes place from the other end, called the FRONT end.

- A linked list is a sequence of nodes in which each node contains one or more than one data field and an address field that stores the address of the next node.

- A stack is a linear collection of data elements in which insertion and deletion take place only at one end, called the TOP of the stack. A stack is a Last-In-First-Out (LIFO) data structure because the last element added to the top of the stack will be the first element to be deleted from the top of the stack.

- A tree is a non-linear data structure in which the data elements or the nodes are represented in a hierarchical order. Here, an initial node is designated as the root node of the tree, and the remaining nodes are partitioned into two disjointed sets such that each set is a part of a sub-tree.

- A binary tree is the simplest form of a tree. A binary tree consists of a root node and two sub-trees known as the left sub-tree and right sub-tree, where both the sub-trees are also binary trees.

- A graph is a general tree with no parent-child relationship. It is a non-linear data structure that consists of vertices or nodes and the edges that connect those vertices.

- An algorithm is a systematic set of instructions combined to solve a complex problem. It is a step-by-finite-step sequence of instructions, each of which has a clear meaning and can be executed with a minimum amount of effort in a finite amount of time.

- The process of dividing an algorithm into modules is called modularization.

- The time complexity of an algorithm is described as the amount of time taken by an algorithm to run the program completely. It is the runtime of the program.

- The space complexity of an algorithm is the amount of memory space required to run the program completely.

- An ADT (Abstract Data Type) is a mathematical model of the data objects that defines a data type as well as the functions to operate on these objects.

- Big O notation is a popular analysis characterization scheme that provides an upper bound on the complexity of an algorithm.

1.10 EXERCISES

Q1. What is a "good" program?

Q2. Explain the classification of data structures.

Q3. What is an algorithm? Discuss the characteristics of an algorithm.

Q4. What are the various operations that can be performed on the data structures? Explain each of them with an example.

Q5. Differentiate a list from a linked list.

Q6. Explain the terms time complexity and space complexity.

Q7. Write a short note on graphs.

Q8. What is the process of modularization?

Q9. Differentiate between stacks and queues with examples.

Q10. What is meant by Abstract Data Type (ADT)? Explain in detail.

Q11. Discuss the worst-case, best-case, and average-case time complexity of an algorithm.

Q12. Write a brief note on trees.

Q13. Explain how you can develop an algorithm to solve a complex problem.

Q14. Explain the time-memory trade-off in detail.

1.11 MULTIPLE CHOICE QUESTIONS

Q1. Which of the following data structures is a FIFO data structure?

a. List

b. Stacks

c. Queues

d. Linked List

Q2. How many maximum children can a binary tree have?

a. 0

b. 2

c. 1

d. 3

Q3. Which of the following data structures uses dynamic memory allocation?

   a. Graphs

   b. Linked Lists

   c. Trees

   d. All of these

Q4. In a queue, deletion is always done from the _____

   a. Front end

   b. Rear end

   c. Middle

   d. None of these

Q5. Which data structure is used to represent complex relationships between the nodes?

   a. Linked Lists

   b. Trees

   c. Stacks

   d. Graphs

Q6. Which of the following is an example of a heterogeneous data structure?

   a. List

   b. Structure

   c. Linked list

   d. None of these

Q7. In a stack, insertion and deletion takes place from the _____

   a. Bottom

   b. Middle

   c. Top

   d. All of these

Q8. Which of the following is not part of the Abstract Data Type (ADT) description?

   a. Operations

   b. Data

   c. Both (a) and (b)

   d. None of the above

Q9. Which of the following data structures allows deletion at both ends of the list but insertion at one end only?

   a. Stack

   b. Input Restricted Dequeue

   c. Output Restricted Dequeue

   d. Priority Queue

Q10. Which of the following data structures is a linear type?

   a. Trees

   b. Graphs

   c. Queues

   d. None of the above

Q11. Which one of the following is beneficial when the data is stored and has to be retrieved in reverse order?

   a. Stack

   b. Linked List

   c. Queue

   d. All of the above

Q12. A binary search tree whose left and right sub-tree differ in height by 1 at most is a _____

   a. Red Black Tree

   b. M way search tree

   c. AVL Tree

   d. None of the above

Q13. The operation of processing each element in the list is called _____

   a. Traversal

   b. Merging

   c. Inserting

   d. Sorting

Q14. Which of the following are the two primary measures of the efficiency of an algorithm?

   a. Data & Time

   b. Data & Space

   c. Time & Space

   d. Time & Complexity

Q15. Which one of the following cases does not exist/occur in complexity theory?

   a. Average Case

   b. Worst Case

   c. Best Case

   d. Minimal Case

## CHAPTER 2

INTRODUCTION TO PYTHON

2.1 INTRODUCTION

What is Python?

Python is a popular programming language. It was created by Guido van Rossum and released in 1991.

It is used for:

- web development (server-side)
- software development
- mathematics
- system scripting

What can Python do?

- Python can be used on a server to create web applications.
- Python can be used alongside software to create workflows.
- Python can connect to database systems. It can also read and modify files.
- Python can be used to handle big data and perform complex mathematics.
- Python can be used for rapid prototyping or production-ready software development.

Why Python?

- Python works on different platforms (Windows, Mac, Linux, and Raspberry Pi).
- Python has a simple syntax similar to the English language.

- Python has a syntax that allows developers to write programs with fewer lines than some other programming languages.
- Python runs on an interpreter system, meaning that code can be executed as soon as it is written. This means that prototyping can be very quick.
- Python can be treated procedurally, in an object-orientated way, or in a functional way.

Good to Know

- The most recent version, Python 3.8 is used for implementation of various codes discussed in this Book. However, Python 2, although not being updated with anything other than security updates, is still quite popular.
- In this tutorial, Python is written in a text editor. It is possible to write Python in an Integrated Development Environment, such as Thonny, Pycharm, Netbeans, or Eclipse, which is particularly useful when managing larger collections of Python files.

Python Syntax Compared to Other Programming Languages

- Python was designed for readability and has some similarities to the English language with influence from mathematics.
- Python uses new lines to complete a command, as opposed to other programming languages that often use semicolons or parentheses.
- Python relies on indentation(using whitespace) to define scope, such as the scope of loops, functions, and classes. Other programming languages often use curly brackets for this purpose.

2.2 PYTHON AND ITS CHARACTERISTICS

Python is a dynamic, high level, free open source, and interpreted programming language. It supports object-oriented programming as well as procedural-oriented programming. In Python, we don't need to declare the type of variable because it is a dynamically typed language. For example, x=10, where x can be anything such as string or int.

Some of its characteristics are as follows:

1. **Easy to Code:** Python is a high-level programming language. Python is very easy to learn as compared to other languages like C, C#, JavaScript, and Java. It is very easy to code in Python and anybody can learn Python in a few hours or days. It is also a developer-friendly language.

2. **Free and Open Source:** Python is freely available at the official website. It is open-source, which means that the source code is available to the public.

3. **Object-Oriented Language:** One of the key features of Python is that it is an object-oriented language and utilizes concepts such as classes and object encapsulation.

4. **GUI Programming Support:** Graphical user interfaces can be made using a module such as PyQt5, PyQt4, wxPython, or Tk in Python. PyQt5 is the most popular option for creating graphical apps with Python.

5. **High-Level Language:** Python is a high-level language. When we write programs in Python, we do not need to remember the system architecture, nor do we need to manage the memory.

6. **Python is Extensible:** We can add some Python code into C or C++ language programs, and we can compile that code in C/C++.

7. **Python is Portable:** Python is also a portable language. For example, if we have Python code developed for Windows, we can run this code on other platforms, such as Linux, Unix, and Mac, without changing it.

8. **Large Standard Library:** Python has a large standard library with a rich set of modules and functions so you do not have to write your code for every single thing. There are many libraries in Python, such as those for regular expressions, unit-testing, and Web browsers.

2.3 PYTHON OVERVIEW

Python is a high-level, interpreted, interactive, and object-oriented scripting language. Python is designed to be highly readable. It uses English keywords frequently (other languages use punctuation), and it has fewer syntactical constructions than other languages.

- **Python is Interpreted** – Python is processed at runtime by the interpreter. You do not need to compile your program before executing it. This is similar to PERL and PHP.

- **Python is Interactive** – You can sit at a Python prompt and interact with the interpreter directly to write your programs.

- **Python is Object-Oriented** – Python supports the object-oriented style or technique of programming that encapsulates code within objects.

- **Python is a Beginner's Language** – Python is a great language for the beginner-level programmers and supports the development of a wide range of applications, from simple text processing to Web browsers and games.

2.4 TOOLS FOR PYTHON

The Anaconda Python distribution is available for Windows, Linux, and Mac, and it is downloadable here:

https://www.anaconda.com/products/individual

Anaconda is well suited for modules such as numpy and scipy, and if you are a Windows user, Anaconda appears to be a better alternative.

2.5 EASY_INSTALL AND PIP

easy_install and pip are Python package installers that will make your life a lot easier when developing in Python.

Whenever you need to install a Python module (and there are many in this book), use either easy_install or pip with the following syntax:

```
easy_install<module-name>
pip install <module-name>
```

2.6 QUOTATIONS AND COMMENTS IN PYTHON

Python allows single ('), double (") and triple (' " or ") quotes for string literals, provided that they match at the beginning and the end of the string. You can use triple quotes for strings that span multiple lines. The following examples are legal Python strings:

```
word = 'word'
line = "This is a sentence."
para = """This is a paragraph. This paragraph contains
more than one sentence."
```

A string literal begins with the letter "r" (for "raw") and treats everything as a literal character and "escapes" the meaning of meta characters (which are discussed in more detail in Chapter 4), as shown here:

```
a1 = r'\n'
a2 = r'\r'
a3 = r'\t'
print'a1:',a1,'a2:',a2,'a3:',a3
```

The output of the preceding code block is here:

a1: \n a2: \r a3: \t

You can embed a single quote in a pair of double quotes (and vice versa) to display a single quote or a double quote. Another way to accomplish the same result is to precede a single or double quote with a backslash character and enclose both in a pair of double-quotes. The following code block illustrates these techniques:

```
b1 = "'"
    b2 = '"'
    b3 = '\''
    b4 = "\""
    print'b1:',b1,'b2:',b2
    print'b3:',b3,'b4:',b4
```

The output of the preceding code block is here:

```
b1: ' b2: "
b3: ' b4: "
```

A hash sign (#) that is not inside a string literal is the character that indicates the beginning of a comment. Moreover, all characters after the hash sign and up to the physical line ending are part of the comment (and are ignored by the Python interpreter). Consider the following code block:

```
#!/usr/bin/python
# First comment
print "Hello, Python!"; # second comment
```

This will produce the following result:
Hello, Python!

A comment may come after a statement or expression on the same line:

```
name = "Tom Jones" # This is also a comment
```

You can place comments on multiple lines as follows:

```
# This is comment one
# This is comment two
# This is comment three
```

A blank line in Python is a line containing only white space, a comment, or both.

2.7 COMPILING THE PYTHON PROGRAM

Compilation: The source code in Python is saved as a .py file, which is then compiled into a format known as byte code. Byte code is then converted to machine code. After the compilation, the code is stored in .pyc files and is regenerated when the source is updated. This process is known as *compilation*.

Linking: *Linking* is the final phase where all the functions are linked with their definitions, as the linker knows where all these functions are implemented.

How to Run a Program in Python 3

1. First, type your program in the Python 3 Idle compiler (or you can use any Python-supported compiler, like Anaconda, Sublime, Python 3 Idle, or any inline compiler).

```
print("hello world")
```

2. Before running your program, save your file and use the Python extension (.py) after your file name (for example, abc.py).

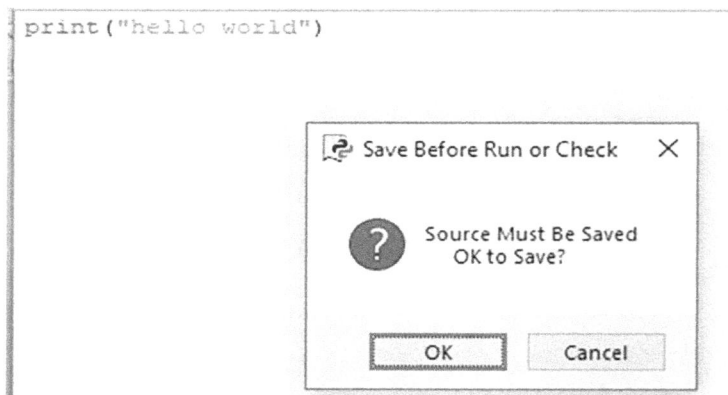

```
print("hello world")
```

```
Save Before Run or Check        ✕

?    Source Must Be Saved
     OK to Save?

       OK              Cancel
```

3. Then press f5 (fn + f5 for Windows) to run your code.

2.8   OBJECT-ORIENTED PROGRAMMING

Object-Oriented Programming (OOP) refers to a type of computer programming (software design) in which programmers define the data type of a data structure and the types of operations (functions) that can be applied to the data structure.

In this way, the data structure becomes an object that includes both data and functions. Programmers can create relationships between one object and another. For example, objects can inherit characteristics from other objects.

The Basic OOP Concepts

If you are new to object-oriented programming languages, you will need to know a few basics before you can get started with code. The following Webopedia definitions will help you better understand object-oriented programming:

- **Abstraction:** The process of picking out (abstracting) common features of objects and procedures
- **Class:** A category of objects. The class defines all the common properties of the different objects that belong to it.
- **Encapsulation:** The process of combining elements to create a new entity. A procedure is a type of encapsulation because it combines a series of computer instructions.
- **Information hiding:** The process of hiding details of an object or function. Information hiding is a powerful programming technique because it reduces complexity.
- **Inheritance:** A feature that represents the "is a" relationship between different classes
- **Interface:** The languages and codes that the applications use to communicate with each other and with the hardware
- **Messaging:** Message passing is a form of communication used in parallel programming and object-oriented programming.
- **Object:** A self-contained entity that consists of both data and procedures to manipulate the data
- **Polymorphism:** A programming language's ability to process objects differently depending on their data type or class.
- **Procedure:** A section of a program that performs a specific task

Advantages of Object-Oriented Programming

One of the principal advantages of object-oriented programming techniques over procedural programming techniques is that they enable programmers to create modules that do not need to be changed when a new type of object is added. A programmer can simply create a new object that inherits many of its features from existing objects. This makes object-oriented programs easier to modify.

OOPL - Object-Oriented Programming Languages

An *object-oriented programming language* (OOPL) is a high-level programming language based on the object-oriented model. To perform object-oriented programming, one needs an object-oriented programming language. Many modern programming languages are object-oriented, however, some older programming languages, such as Pascal, do offer object-oriented versions. Examples of object-oriented programming languages include Java, C++, and Smalltalk.

2.9   CHARACTER SET USED IN PYTHON

The character set allowed in Python consists of the following characters:

1. **Alphabet** – This includes uppercase as well as lowercase letters of English, i.e., {A, B, C... ., Z} and {a, b, c... ., z}.

2. **Digits** – This includes decimal digits, i.e., {0, 1, 2 . . ., 9}.

3. **White Spaces** – This includes spaces, enters, and tabs.

4. **Special Characters** – These consist of special symbols, including {, !, ?, #, <, >, (,), %, ", &, ^, *, <<, >>, [,], +, =, /, -, _, :, ;, }.

2.10 PYTHON TOKENS

1. Keywords
2. Identifiers
3. Literals
4. String
5. Numeric
6. Collection literals
7. List Literals
8. Tuples

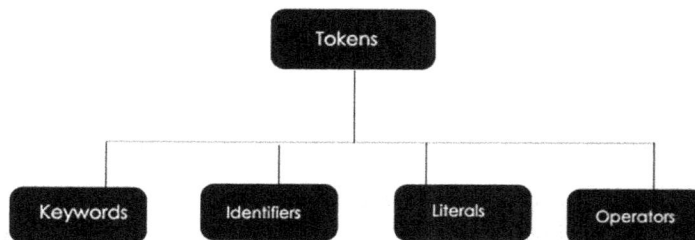

FIGURE 2.1 The relationship between tokens and other entities

Keywords

These are the dedicated words that have special meaning and functions. The compiler defines these words. It does not allow users to use these words.

Identifiers:
Identifiers represent the programmable entities. The programmable entities include user-defined names, variables, modules, and other objects. Moreover, Python has some rules for defining the identifiers. For example, an identifier can be a sequence of lowercase letters, uppercase letters, integers, or a combination of any of those. The identifier name should start with a lower case or upper case letter (it must not start with digits). The identifier name should not be a reserved word. Only the underscore (_) can be used as a special character in identifier names. The length of the identifier name should not be more than 79 characters.

Literals

Literals are used to define the data as a variable or constant. Python has 6 literal tokens.

String

The string is a sequence of characters defined between quotes (Both single and double quotes are applicable to define the string literals.). These strings perform several operations. Let us discuss some of them.

Numeric

These are immutable (unchangeable) literals. We have 3 different numerical types: integer, float, and complex.

Boolean

This has only two values: true or false.

Collection literals

A collection literal is a syntactic expression form that evaluates an aggregate type, such as an array list or map. Python supports 2 types of collection literal tokens.

List Literals

You can consider Python lists as arrays in C. But the difference between the arrays and lists is that arrays hold the homogeneous data type and lists hold the heterogeneous data types. The list is the most versatile data type in Python. Python literals are separated by a comma in [].

Note: If a comma is not provided between the values, the output does not contain spaces.

Example

```
List = ['a','b','c']
Print(list)
Output :
['a','b','c']
```

Tuples

Tuples are similar to lists. But like lists, tuples cannot change values. Tuples are enclosed in parentheses, whereas lists are enclosed in square brackets. Tuples perform all the same operations as lists do.

Set

A set is a well-defined collection of elements. The elements in the set are placed in curly braces separated by a comma. In the set, every element is unique.

Set 1 = {1, 2, 3}
Set 2 = {1, 2, 2, 3}

In the above example, element 2 is taken twice. Now, let us discuss the various set operations.

Union

This combines all the elements in the string. The union operation is performed using the pipe (|) operator tokens.

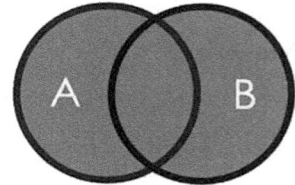

Example

A = {1, 2, 3, 4, 5, 6}
B = {3, 4, 5, 6, 7, 8}
A|B = {1, 2, 3, 4, 5, 6, 7, 8}

Intersection

The intersection of A and B returns the common elements in the sets. The operation is performed using the and operator tokens.

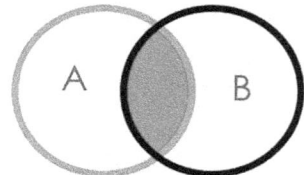

Example

A = {1, 2, 3, 4, 5, 6}
B = {3, 4, 5, 6, 7, 8}
A & B = {3, 4, 5, 6}

Difference

The difference of (A–B) returns the elements that are only in A, but not in B. Similarly, B–A returns only the elements that are only in B but not in A tokens.

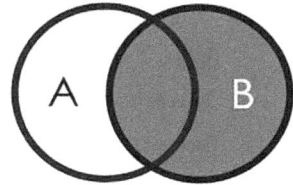

Example

A = {1, 2, 3, 4, 5, 6}
B = {3, 4, 5, 6, 7, 8}
A–B = {1, 2}
B–A = {7, 8}

Symmetric difference

This returns the set of elements that are in both A and B, except for the common element tokens.

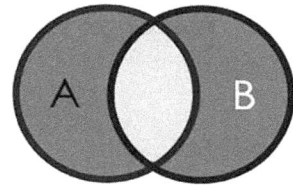

Example

A = {1, 2, 3, 4, 5, 6}
B = {3, 4, 5, 6, 7, 8}
A^B = {1, 2, 7, 8}

Dictionaries

Python dictionaries are the key value pairs that are enclosed in curly braces. Dictionaries are separated by the ":"

```
Dict = {'name' : 'Onlineitguru', age : 20}
```

These elements are accessed as `Dict['name']`.

Output: Onlineitguru

Appending the elements in dictionaries is written as follows:

```
Dict['address']=ameerpet
```

Output

'name'='onlineitguru','age'=20, address'=ameerpet.

2.11 DATA TYPES IN PYTHON

Data types are the special keywords that define the type of data and the amount of data a variable is holding. In programming, a data type is an important concept. Variables can store data of different types, and different types can do different things.

Python has the following data types built-in by default, in these categories:

Text Type:	str
Numeric Types:	int, float, complex
Sequence Types:	list, tuple, range
Mapping Type:	dict
Set Types:	set, frozenset
Boolean Type:	bool
Binary Types:	bytes, bytearray, memoryview

Getting the Data Type

You can get the data type of any object by using the type() function.

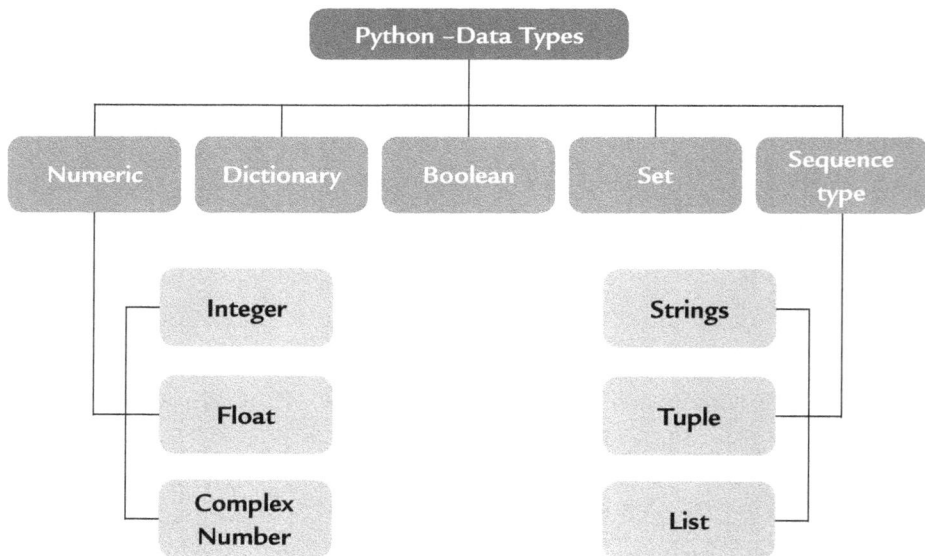

FIGURE 2.2 The relationship between Python data types

2.12 STRUCTURE OF A PYTHON PROGRAM

In general, a Python program consists of many text files that contain Python statements. A program is designed as a single main, high-level file with one or more supplemental files.

In Python, a high-level file has an important path of control that you must follow before you can start your application. The library tools are also known as *module files*. These tools are implemented for making a collection of top-level files. High-level files use tools that are defined in module files. Module files implement files that are defined in other modules. In Python, a file takes a module to get access to the tools it defines. The tools are made by a module's type. The final thing we take are the modules and access the attributes of their tools. This shows the programming structure of Python.

Attributes and Imports:

The structure of a Python program consists of three files, such as a.py, b.py, and c.py. The file model a.py is used for a high-level file. It is known as a simple text file of statements. It can be executed from bottom to top when it is launched. Files b.py and c.py are modules. They are considered text files of statements, but they are generally not started directly. These attributes define the programming structure of Python.

Functions:

For example, b.py defines a function called spam. For external use, b.py has a Python def statement to start the function. It is later operated by passing one or more values, as shown in the following example.

```
Def  spam(text): print text, 'spam'
```
If a.py wants to use spam, it uses the following Python statement:
```
Import b b.spam ('gumby')
```

Statements:

The Python import statement gives the file a.py access to file b.py. It shows "load fileb.py" and gives access to all its attributes named "b." The import statements will execute and implement another file at runtime. The cross-file module is not updated until the import statements are executed.

The next part is the statements that call the function spam. Module b uses object attribute notation. B.spam means to get the value of name spam within object b. We can implement a string in parentheses if these files are run by a.py.

In regular usage, we see Object. Attribute in Python scripts. Many objects have attributes traced by Python operators.

The process of importing is considered common in Python. Any sort of file can get tools from any file. Chains can go as deep as you want. You can get notified module a to import b and b can import c, and c again imports b.

Modules:

One way to view this is to see Python as a big company structure. Modules have the top end of the code. The coding components in module files are used in the program files. Let's take an example function: b.spam is a regular purpose tool. We can again implement that in a different program. This is simply known as b.py from any other program files.

Standard library files:

> Python has a large collection of modules known as the standard library. It contains 200 modules (at last count). It includes platform-independent common programming things, such as GUI design, Internet and network scripting, text design matching, and operating system interfaces. All of these are used in the programming structure of Python.

2.13 OPERATORS IN PYTHON

> Operators are used for performing operations on variables and values.
> Python divides the operators into the following groups:

- Arithmetic operators
- Assignment operators
- Comparison operators
- Logical operators
- Identity operators
- Membership operators
- Bitwise operators

Python Arithmetic Operators

> Arithmetic operators are used with numeric values to perform common mathematical operations.

Table 2.1 Common mathematical operations

Operator	Name	Example
+	Addition	x + y
−	Subtraction	x − y
*	Multiplication	x * y
/	Division	x / y
%	Modulus	x % y
**	Exponentiation	x ** y
//	Floor division	x // y

Python Assignment Operators

Assignment operators are used for assigning values to variables.

Table 2.2 Assignment operators

Operator	Example	Same As
=	x = 5	x = 5
+=	x += 3	x = x + 3
–=	x –= 3	x = x – 3
*=	x *= 3	x = x * 3
/=	x /= 3	x = x / 3
%=	x %= 3	x = x % 3
//=	x //= 3	x = x // 3
**=	x **= 3	x = x ** 3
&=	x &= 3	x = x & 3
|=	x |= 3	x = x | 3
^=	x ^= 3	x = x ^ 3
>>=	x >>= 3	x = x >> 3
<<=	x <<= 3	x = x << 3

Python Comparison Operators

Comparison operators are used for comparing two values.

Table 2.3 Comparison operators

Operator	Name	Example
==	Equal	x == y
!=	Not equal	x != y
>	Greater than	x > y
<	Less than	x < y
>=	Greater than or equal to	x >= y
<=	Less than or equal to	x <= y

Python Logical Operators

Logical operators are used for combining conditional statements.

Table 2.4 Logical operators

Operator	Description	Example	Try it
and	Returns True if both statements are true	x < 5 and x < 10	Try it »
or	Returns True if one of the statements is true	x < 5 or x < 4	Try it »
not	Reverse the result, returns False if the result is true	not(x < 5 and x < 10)	Try it »

Python Identity Operators

Identity operators are used for comparing the objects, not if they are equal, but if they are actually the same object, with in the same memory location.

Table 2.5 Identity operators

Operator	Description	Example	Try it
is	Returns True if both variables are the same object	x is y	Try it »
is not	Returns True if both variables are not the same object	x is not y	Try it »

Python Membership Operators

Membership operators are used for testing if a sequence is presented in an object.

Table 2.6 Membership operators

Operator	Description	Example	Try it
in	Returns True if a sequence with the specified value is present in the object	x in y	Try it »
not in	Returns True if a sequence with the specified value is not present in the object	x not in y	Try it »

Python Bitwise Operators

Bitwise operators are used for comparing (binary) numbers.

Table 2.7 Bitwise operators

Operator	Name	Description
&	AND	Sets each bit to 1 if both bits are 1
|	OR	Sets each bit to 1 if one of two bits is 1
^	XOR	Sets each bit to 1 if only one of two bits is 1
~	NOT	Inverts all the bits
<<	Zero fill left shift	Shift left by pushing zeros in from the right and let the left-most bits fall off
>>	Signed right shift	Shift right by pushing copies of the left-most bit in from the left, and let the right-most bits fall off

PROGRAM TO ADD TWO NUMBERS ACCEPTED BY USER

main.py   ▷ Run

```
1   # Store input numbers
2   num1 = input('Enter first number: ')
3   num2 = input('Enter second number: ')
4
5   # Add two numbers
6   sum = float(num1) + float(num2)
7
8   # Display the sum
9   print('The sum of {0} and {1} is {2}'.format(num1, num2, sum))
10
```

Shell

```
Enter first number: 58
Enter second number: 97
The sum of 58 and 97 is 155.0
>>>
```

2.14 DECISION CONTROL STATEMENTS

The *Decision Control Statement* (DCS) is a statement that determines the control flow of a set of instructions. This means the DCS decides the sequence in which the instructions in the program are executed.

The three fundamental methods of control flow in a programming language are

1. Sequential Control

2. Selection Control

3. Iterative Control

Here, we only discuss two methods of control flow.

Sequential Control

Sequential control is when the program is executed line by line, meaning from the first line to the second line then from the second line to the third line and so on.

Selection Control Statement

When we execute only a selected set of statements, then we use the Selection Control Statement. It usually jumps from one part of the code to another depending on whether a particular condition is satisfied or not.

In the Selection Control Statement, we need

- If statement
- If-else statement
- If-elif-else statement

If Statement

The if statement is the simplest form of decision control statement that is frequently used in decision making.

Syntax of the If Statement

```
if (test_expression) :
statement1
............. .
statement n
statement x
```

Program to increment a number if it is positive

```
x = 10 # Initialize the value of x
if (x>0): # test the value of x
x = x+1 # Increment the value of x if it is > 0
print(x) # print the value of
```

If-else Statement

The use of If-else statement is very simple. When you run your program, the test expression is evaluated. If the result is True, the statement followed by the expression is executed, else, if the expression is False, the statement followed by the expression is executed.

Syntax of the If-else Statement

```
if ( test expression ) :
statement block 1
else :
statement block 2
statement x
```

PROGRAM TO FIND WHETHER THE NUMBER IS EVEN OR ODD

main.py   ▷ Run

```
1   # Python program to check if the input number is odd or even.
2   # A number is even if division by 2 gives a remainder of 0.
3   # If the remainder is 1, it is an odd number.
4
5   num = int(input("Enter a number: "))
6 ▾ if (num % 2) == 0:
7       print("{0} is Even".format(num))
8 ▾ else:
9       print("{0} is Odd".format(num))
```

Shell

```
Enter a number: 331
331 is Odd
>>> |
```

Shell

```
Enter a number: 44
44 is Even
>>>
```

If-elif-else statement

Python supports if-elif-else statements to test additional conditions apart from the initial test expression. The if-elif-else constructs works in the same way as usual to if-else statement. One more thing to remember that it is not necessary that every if statement should have an else block, as Python supports simple if statements, also.

Syntax of If-elif-else statement

```
if ( test expression 1):
statement block 1
elif ( test expression 2 ):
statement block 2
.............................
elif ( test expression N ):
statement block N
else:
statement block X
Statement Y
```

2.15 LOOPING STATEMENTS

In general, statements are executed sequentially: The first statement in a function is executed first, followed by the second, and so on. There may be a situation when you need to execute a block of code several times.

Programming languages provide various control structures that allow for more complicated execution paths.

A loop statement allows us to execute a statement or group of statements multiple times. The following diagram illustrates a loop statement.

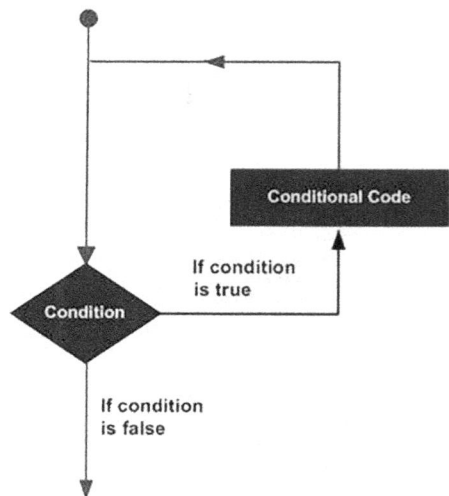

The Python programming language provides the following types of loops to handle looping requirements.

Table 2.8 Types of loops

Sr. No.	Loop type and description
1	While loop Repeats a statement or group of statements while a given condition is TRUE. It tests the condition before executing the loop body.
2	For loop Executes a sequence of statements multiple times and abbreviates the code that manages the loop variable.
3	Nested loops You can use one or more loops inside any another while, for, or do.. while loop.

A sample while loop program is shown in the graphic.

```
i = 1
while i < 6:
    print(i)
    i += 1
```

OUTPUT

```
1 2 3 4 5
```

A for loop program is shown in the following graphic.

```python
fruits = ["apple", "banana", "cherry"]
for x in fruits:
    print(x)
```

OUTPUT

```
apple
banana
cherry
```

A nested loop program is as follows.

```python
i = 2
while(i < 100):
    j = 2
    while(j <= (i/j)):
        if not(i%j): break
        j = j + 1
    if (j > i/j) : print i, " is prime"
    i = i + 1

print "Good bye!"
```

OUTPUT

```
2 is prime
3 is prime
5 is prime
7 is prime
11 is prime
13 is prime
17 is prime
19 is prime
23 is prime
29 is prime
31 is prime
37 is prime
41 is prime
43 is prime
47 is prime
53 is prime
59 is prime
61 is prime
67 is prime
71 is prime
73 is prime
79 is prime
83 is prime
89 is prime
97 is prime
Good bye!
```

2.16 LOOP CONTROL STATEMENTS

Loop control statements change the execution from its normal sequence. When the execution leaves a scope, all automatic objects that were created in that scope are destroyed.

Python supports the following control statements. Let us go through the loop control statements briefly.

Table 2.9 Examples of loop control statements

Sr. No.	Control Statement & Description
1	Break statement Terminates the loop statement and transfers execution to the statement immediately following the loop.
2	continue statement It causes the loop to skip the remainder of its body and immediately retest its condition prior to reiterating.
3	pass statement The pass statement in Python is used when a statement is required syntactically, but you do not want any command or code to execute.

An example of a break statement program is shown below.

main.py   ▷ Run

```
1   # Use of break statement inside the loop
2
3 ▾ for val in "string":
4 ▾     if val == "i":
5           break
6       print(val)
7
8   print("The end")
```

OUTPUT

```
Shell

s
t
r
The end
>>>
```

2.17 METHODS

Introduction to Python Methods

You are aware of the fact that Python is an object-oriented language, right? This means that it can deal with classes and objects to model the real world. A Python method is a label that you can call on an object; it is a piece of code to execute on that object.

Python Class Method

A Python class is an Abstract Data Type (ADT). Think of it as a blueprint. A rocket is made by referring to its blueprint, that is, according to its plan. It has all the properties mentioned in the plan and behaves accordingly. Likewise, a class is a blueprint for an object. For example, consider a car. The class "Car" contains properties like brand, model, color, and fuel. It also holds behavior like start(), halt(), drift(), speedup(), and turn(). An object Hyundai Verna has the following properties:

brand: 'Hyundai'
model: 'Verna'
color: 'Black'
fuel: 'Diesel'

Here, this is an object of the class Car, and we may choose to call it "car1" or "blackverna."

```
1.  >>> class Car:
2.  def__init__(self,brand,model,color,fuel):
3.  self.brand=brand
4.  self.model=model
5.  self.color=color
6.  self.fuel=fuel
7.  defstart(self):
8.  pass
9.  defhalt(self):
10. pass
11. defdrift(self):
12. pass
13. defspeedup(self):
14. pass
15. defturn(self):
16. pass
```

Python Objects

A Python object is an instance of a class. It can have properties and behavior. We just created the class Car. Now, let's create an object blackverna from this class. Remember that you can use a class to create as many objects as you want.

```
1.  >>>blackverna=Car('Hyundai','Verna','Black','Diesel')
```

This creates a Car object, called blackverna, with the aforementioned attributes. We did this by calling the class like a function (the syntax). Now, let's access its fuel attribute. To do this, we use the dot operator in Python(.).

```
2.  >>>blackverna.fuel
```

The output is "Diesel."

Python Method

A Python method is like a Python function, but it must be called on an object. To create it, you must put it inside a class. Now in this Car class, we have five methods, namely, start(), halt(), drift(), speedup(), and turn(). In this example, we put the pass statement in each of these, because we haven't decided what to do yet. Let's call the drift() Python method on blackverna.

```
1.  >>>blackverna.drift()
2.  >>>
```

Like a function, a method has a name and may take parameters and have a return statement. Let's take an example of this.

```
1.  >>> class Try:
2.  def__init__(self):
3.  pass
4.  defprinthello(self,name):
5.  print(f"Hello, {name}")
6.  return name
7.  >>>obj=Try()
8.  >>>obj.printhello('Ayushi')
```

The output is

Hello, Ayushi
'Ayushi'

Here, the method printhello() has a name, takes a parameter, and returns a value.

An interesting discovery – When we first defined the class Car, we did not pass the self parameter to the five methods of the class. This worked fine with the attributes, but when we called the drift() method on blackverna, it gave us this error:

Traceback (most recent call last):
File "<pyshell#19>", line 1, in <module>
blackverna.drift()
TypeError: drift() takes 0 positional arguments but 1 was given

From this error, we figured that we were missing the self parameter to all those methods. Then we added it to all of them and called drift() on blackverna again. It still didn't work.

Finally, we declared the blackverna object again, and then called drift() on it. This time, it worked without an issue. Make out of this information what you will.

__init__()

If you're familiar with any other object-oriented language, you know about constructors. In C++, a constructor is a special function, with the same name as the class, used to initialize the class's attributes. Here in Python, __init__() is the method we use for this purpose. Let's see the __init__ part of another class.

```
1.  >>> class Animal:
2.  def__init__(self,species,gender):
3.  self.species=species
4.  self.gender=gender
5.  >>> fluffy=Animal('Dog','Female')
6.  >>>fluffy.gender
```

The output is "Female."
Here, we used __init__ to initialize the attributes species and gender.
However, you don't need to define this function if you don't need it in your code.

```
1.  >>> class Try2:
2.  defhello(self):
3.  print("Hello")
4.  >>> obj2=Try2()
5.  >>> obj2.hello()
```

Init is a magic method, which is why it has double underscores before and after it. We will learn about magic methods in a later section in this article.

Python Self Parameter

You would have noticed until now that we've been using the self-parameter with every method, even the __init__(). This tells the interpreter to deal with the current object. It is like the "this" keyword in Java. Let's take another code to see how this works.

```
1.  >>> class Fruit:
2.  def printstate(self, state):
3.  print(f"The orange is {state}")
4.  >>> orange=Fruit()
5.  >>>orange.printstate("ripe")
```

The output is as follows:
The orange is ripe.

As you can see, the self-parameter told the method to operate on the current object, that is, orange. Let's take another example.

```
1.  >>> class Result:
2.  def __init__(self, phy, chem, math):
3.  self.phy=phy
4.  self.chem=chem
5.  self.math=math
6.  def printavg(self):
7.  print(f"Average={(self.phy+self.chem+self.math)/3}")
8.  >>>rollone=Result(86,95,85)
9.  >>>rollone.chem
```

The output is as follows: 95.

```
1.  >>>rollone.printavg()
```

Average=88.66666666666667
You can also assign values directly to the attributes, instead of relying on arguments.

```
1.  >>> class LED:
2.  def __init__(self):
3.  self.lit=False
4.  >>>obj=LED()
5.  >>>obj.lit
```

The output is as follows:
False

Finally, we'd like to say that self isn't a keyword. You can use any name instead of it, provided that it isn't a reserved keyword, and follows the rules for naming an identifier.

```
1.  >>> class Try3:
2.  def __init__(thisobj,name):
3.  thisobj.name=name
4.  >>> obj1=Try3('Leo')
5.  >>> obj1.name
```

The output is as follows:
Leo

Python Functions vs. Method

A function differs from a method in the following ways.

1. While a method is called on an object, a function is generic.

2. Since we call a method on an object, it is associated with it. Consequently, it is able to access and operate on the data within the class.

3. A method may alter the state of the object; a function does not when an object is passed as an argument to it. We have seen this in our tutorial on tuples.

Python Magic Methods

Another construct that Python provides us with is Python magic methods. Such a method is identified by double underscores before and after its name. Another name for a magic method is a dunder.

A magic method is used to implement functionality that can't be represented as a normal method. __init__() isn't the only magic method in Python. But for now, we'll just name some of the magic methods:

```
__add__  for +
__sub__  for -
__mul__  for *
__and__  for &
```

The list, however, does not end here.

Conclusion: Python Method

A Python method, as we know it, is much like a function, except for the fact that it is associated with an object. Now you know how to define a method, and make use of the __init__ method and the self-parameter, or whatever you choose to call it. Don't forget to revise the various methods we discussed in our tutorials on Python lists, tuples, strings, sets, and dictionaries in Python.

2.18 SUMMARY

- Python is a popular programming language. It was created by Guido van Rossum and released in 1991.

- Python is a dynamic, high level, free, open source, and interpreted programming language. It supports object-oriented programming as well as procedural oriented programming.

- Object-oriented programming (OOP) refers to a type of computer programming (software design) in which programmers define the data type of a data structure and the types of operations (functions) that can be applied to the data structure.

- **Class:** A category of objects. The class defines all the common properties of the different objects that belong to it.

- **Encapsulation:** The process of combining elements to create a new entity. A procedure is a type of encapsulation because it combines a series of computer instructions.

- **Inheritance:** A feature that represents the "is a" relationship between different classes

- **Object:** A self-contained entity that consists of both data and procedures to manipulate the data

- Data types are the special keywords that define the type of data and the amount of data a variable is holding.

- Operators are used for performing operations on variables and values.

- Arithmetic operators are used with numeric values to perform common mathematical operations.

- Assignment operators are used for assigning values to variables.

- Comparison operators are used for comparing two values.

- Logical operators are used for combining conditional statements.

- Identity operators are used for comparing the objects, not if they are equal, but if they are actually the same object, with the same memory location.

- Membership operators are used for testing if a sequence is presented in an object

- Bitwise operators are used for comparing (binary) numbers.

- The Decision Control Statement is a statement that determines the control flow of a set of instructions. The DCS decides the sequence in which instructions in the program are to be executed.

- The if statement is the simplest form of the decision control statement that is frequently used in decision making.

- The use of if -else statement is very simple. When you run your program, the test expression is evaluated. If the result is True, the statement followed by the expression is executed, else if the expression is False, the statement followed by the expression is executed.

- Python supports if-elif-else statements to test additional conditions apart from the initial test expression. The if-elif-else constructs works in the same way as the if-else statement.

- **While loop-**Repeats a statement or group of statements while a given condition is TRUE. It tests the condition before executing the loop body.

- **For loop-**Executes a sequence of statements multiple times and abbreviates the code that manages the loop variable.

- **Nested loop-**You can use one or more loops inside any another while, for, or do..while loop.

- **Break statement-**Terminates the loop statement and transfers execution to the statement immediately following the loop

- **Continue statement-**Causes the loop to skip the remainder of its body and immediately retest its condition prior to reiterating

- **Pass statement-**The pass statement in Python is used when a statement is required syntactically, but you do not want any command or code to execute.

- A Python Class is an Abstract Data Type (ADT).

- A Python object is an instance of a class.

- A Python method is like a Python function, but it must be called on an object. To create it, you must put it inside a class.

- A magic method is used to implement functionality that can't be represented as a normal method.

2.19 EXERCISES

2.19.1 Theory Questions

Q1. What is the difference between lists and tuples in Python?

Q2. What are the key features of Python?

Q3. What type of language is Python?

Q4. How is Python an interpreted language?

Q5. What is pep 8?

Q6. How is memory managed in Python?

Q7. What is name space in Python?

Q8. What is Python path?

Q9. What are Python modules?

Q10. What are local variables and global variables in Python?

Q11. Is Python case sensitive?

Q12. What is type conversion in Python?

Q13. How can you install Python on Windows and set the path variable?

Q14. Is indentation required in Python?

Q15. What is the difference between Python arrays and lists?

Q16. What are functions in Python?

Q17. What is __init__?

Q18. What is a lambda function?

Q19. What is self in Python?

Q20. How do break, continue, and pass work?

Q21. What does [::-1} do?

Q22. How can you randomize the items of a list in place in Python?

2.19.2 Programming Projects

Write the following programs:

Q1. Python program to add two numbers

Q2. Python program for the factorial of a number

Q3. Python program for simple interest

Q4. Python program for compound interest

Q5. Python program to check the Armstrong Number

Q6. Python program to find the area of a circle

Q7. Python program to print all prime numbers in an interval

Q8. Python program to check whether a number is prime or not

Q9. Python program for nth Fibonacci number

Q10. Python program for Fibonacci numbers

Q11. Python program to check if a given number is a Fibonacci number

Q12. Python program for nth multiple of a number in a Fibonacci Series

Q13. Program to print the ASCII Value of a character

Q14. Python program for the sum of squares of the first n natural numbers

Q15. Python program for the cube sum of the first n natural numbers

2.19.3 Multiple Choice Questions

Q1. What will be the output of the following Python code?

```
print("Hello {name1} and {name2}".format(name1='foo',
name2='bin'))
```

a. Hello foo and bin

b. Hello {name1} and {name2}

c. Error

d. Hello and

Q2. What will be the output of the following Python code?

```
print("Hello {0!r} and {0!s}".format('foo','bin'))
```

a. Hello foo and foo

b. Hello 'foo' and foo

c. Hello foo and 'bin'

d. Error

Q3. What will be the output of the following Python code?

```
print("Hello {0} and {1}".format(('foo','bin')))
```

a. Hello foo and bin

b. Hello ('foo', 'bin') and ('foo', 'bin')

c. Error

d. None of the mentioned

Q4. What will be the output of the following Python code?

```
print("Hello {0[0]} and {0[1]}".format(('foo','bin')))
```

a. Hello foo and bin

b. Hello ('foo', 'bin') and ('foo', 'bin')

c. Error

d. None of the mentioned

Q5. What will be the output of the following Python code snippet?

```python
print('The sum of {0} and {1} is {2}'.format(2,10,12))
```

a. The sum of 2 and 10 is 12

b. Error

c. The sum of 0 and 1 is 2

d. None of the mentioned

Q6. What will be the output of the following Python code snippet?

```python
print('The sum of {0:b} and {1:x} is {2:o}'.format(2,10,12))
```

a. The sum of 2 and 10 is 12

b. The sum of 10 and a is 14

c. The sum of 10 and a is c

d. Error

Q7. What will be the output of the following Python code snippet?

```python
print('{:,}'.format(1112223334))
```

a. 1,112,223,334   **b.** 111,222,333,4

c. 1112223334   **d.** Error

Q8. What will be the output of the following Python code snippet?

```python
print('{:,}'.format('1112223334'))
```

a. 1,112,223,334   **b.** 111,222,333,4

c. 1112223334   **d.** Error

Q9. What will be the output of the following Python code snippet?

```python
print('{:$}'.format(1112223334))
```

a. 1,112,223,334   **b.** 111,222,333,4

c. 1112223334   **d.** Error

Q10. What will be the output of the following Python code snippet?

```python
print('{:#}'.format(1112223334))
```

a. 1,112,223,334   **b.** 111,222,333,4

c. 1112223334   **d.** Error

## CHAPTER 3

ARRAYS/LISTS

3.1 INTRODUCTION

We studied the basics of programming using data structures and Python in the previous chapter. We discussed how to design good programs that run correctly and efficiently by occupying less space in the memory and take little time to run and execute. A program is said to be efficient when it executes with little memory space and in a minimal amount of time. In this chapter, we discuss the concept of arrays. An array is a user-defined data type that stores related information together. In Python, a list is similar to an array. But there are two major differences between the array and the list. First, an array has a limited number of operations, which commonly include those for array creation, reading a value from a specific element, and writing a value to a specific element. The list, on the other hand, provides a large number of operations for working with the contents of the list. Second, the list can grow and shrink during execution as elements are added or removed, while the size of an array cannot be changed after it has been created.

3.2 DEFINITION OF AN ARRAY

An *array* is a collection of homogeneous (similar) types of data elements in contiguous memory. An array is a linear data structure where all the elements of the array are stored in linear order. Let us take an example in which we have ten students in a class. We have been asked to store the grades of all ten students; we need to use an array.

36	98	14	74	56	13	7	96	44	82
grades1	grades2	grades3	grades4	grades5	grades6	grades7	grades8	grades9	grades10

FIGURE 3.1 Representation of an array of 10 elements

In the previous example, the data elements are stored in the successive memory locations and are identified by an index number (also known as the subscript), that is, A_i or A[i]. A *subscript* is an ordinal number used to identify an element of the array. The elements of an array have the same data type, and each element in an array can be accessed using the same name.

> **Frequently Asked Questions**
>
> **Q. What is an array? How can we identify an element in the array?**
>
> **Answer:**
>
> *An array is a collection of homogeneous (similar) types of data elements in contiguous memory. An element in an array can be identified by its index number, which is also known as a subscript.*

3.3 ARRAY/LIST DECLARATION

We know that all variables must be declared before they are used in the program. But in python, it is not mandatory to declare variables before use. Declaring an array involves the following specifications:

- *Data Type* – The data type means the different kinds of values it can store. The data type can be an integer(int), float, char, or any other valid data type.
- *Array Name* – The name refers to the name of the array which will be used to identify the array.
- *Size* – The size of an array refers to the maximum number of values an array can hold.

Syntax –

```
List=[]
```

Example:

```
Salary=[]
```

The previous example declares the salary to be an array. In Python, the array index starts from zero. The first element of this array will be stored in salary [0], the second element will be stored in salary [1], and so on. In memory, the array is arranged as shown in Figure 3.2.

2000	4500	7890	9876	10000	3458	8000	9810	14000	5000
salary[0]	salary[1]	salary[2]	salary[3]	salary[4]	salary[5]	salary[6]	salary[7]	salary[8]	salary[9]

FIGURE 3.2 Memory representation of an array

Here the values 0, 1, 2, . . .9 written in square brackets represent the subscripts we use to identify a particular element in the array.

3.4 ARRAY/LIST INITIALIZATION

The initialization of arrays can be done with the initialization at the compile time. The initialization of the elements of the array at compile time refers to the same way we initialize the normal or ordinary variables at the time of their declaration. When an array is initialized, there is a need to provide a specific value for every element in the array.

The general form of initializing arrays is as follows:

```
array_name = [list of values]
```

An example of the initialization of arrays at compile time is as follows.

During the initialization of arrays, we may omit the size of the array. For example,

```
age = [20, 25, 23, 28, 30]
```

In the previous example, the compiler will automatically allocate memory for all the initialized elements of the array. For example,

```
marks = [56, 69, 40,99, 82, 96, 72]
```

56	69	40	99	82	96	72
grades[0]	grades[1]	grades[2]	grades[3]	grades[4]	grades[5]	grades[6]

FIGURE 3.3 Initialization of the array grades

3.5 CALCULATING THE ADDRESS OF ARRAY ELEMENTS

The address of the elements in the 1-D array can be calculated very easily, because the array stores all its data elements in contiguous memory locations, storing the base address (address of the first element of the array). Hence, the address of the other data elements can easily be calculated using the base address. The formula to find the address of elements in a 1-D array is as follows:

Address of data element, A[i] = Base Address (BA) + w (i − lower bound)

where A is the array, i is the index of the element for which the address is to be calculated, BA is the base address of the array A, and w is the size of each element (e.g., the size of int is 2 bytes, the size of char is 1 byte.)

Frequently Asked Questions

Q. An array is given the grades [34, 53, 87, 100, 98, 65]. **Calculate the address of** *grades[3]* **if the base address is 3000.**

Answer:

The base address of the array is 3000 and we know that the size of an integer is 2 bytes. Hence, we can easily find the address of **grades[3]** *by putting the information into the formula:*

3000	3002	3004	3006	3008	3010
34	53	87	100	98	65
grades[1]	grades[2]	grades[3]	grades[4]	grades[5]	grades[6]

Address of grades[3] = 3000 + 2 (3 − 1)

$$= 3000 + 2 \ (2)$$

Address of grades[3] = 3004

3.6 OPERATIONS ON ARRAYS/LISTS

This section discusses various operations that can be performed on arrays/lists. These operations include

- Traversing an array/list
- Inserting an element into an array/list
- Deleting an element in an array/list
- Searching for an element in an array/list
- Merging two arrays/lists
- Sorting arrays/lists

1. Traversing an Array/List

Traversing an array/list means to access every element in an array/list exactly once so that it can be processed. Examples are counting all the data elements or performing any process on these elements. Traversing the elements of the array/list is a very simple process because of the linear structure of the array/list (all the elements are stored in the contiguous memory locations).

Practical Application

Imagine there is a line of people standing one behind the other. One boy is distributing advertisement pamphlets one by one to each person standing in the line.

Code for Traversing an Array/List:

```
# Python 3 code to iterate over a list
list = [1, 3, 5, 7, 9]

# Using for loop
for i in list:
    print(i)
```

2. Inserting an Element in an Array/List

append()	Adds an element at the end of the list
insert()	Adds an element at the specified position

For example:

Numbers=[12,11,23,44]

1. append( )

Code:

```
Numbers.append(55)
```

Output:

12,11,23,44,55

2. insert( )

Code:

```
Numbers.append(2,47)
```

Output:

12,11,47,44,55

3. Deleting an Element in an Array/List

pop()	Removes the element at the specified position
remove()	Removes the first item with the specified value

For example:

```
Numbers=[12,11,23,44]
1. pop( )
```

Code:

```
Numbers.pop(2)
```

Output:

```
12,11,44,55
2. remove( )
```

Code:

```
Numbers.remove( )
```

Output:

11,47,44,55

Code:

```
Numbers.remove(47 )
```

Output:

12,11,44,55

4. Searching for an Element in an Array/List

index()	Returns the index of the first element with the specified value

For example:

```
Numbers=[12,11,23,44]
index( )
```

Code:

```
Numbers.index(23)
```

Output:

2

5. Merging of Two Array/Lists

We can merge two array/lists in Python by simply adding them.

For example:

Code:

```
listone = [1,2,3]
listtwo = [4,5,6]
joinedlist = listone + listtwo
```

Output:

1,2,3,4,5,6

6. Sorting an Array/List

sort()	Sorts the list

For Example:

List= [12,33,11,77,43,55]

Code:

```
List.sort( )
```

Output:

[11,12,33,43,55,77]

3.7 2-D ARRAYS/TWO-DIMENSIONAL ARRAYS

We have already discussed one-dimensional arrays and their various types and operations. Now, we discuss two-dimensional arrays. Unlike one-dimensional arrays, 2-D arrays are organized in the form of grids or tables. They are a collection of 1-D arrays. One-dimensional arrays are organized linearly in the memory. A 2-D array consists of two subscripts:

1. First subscript – which denotes the row

2. Second subscript – which denotes the column

A 2-D array is represented as shown in the following figure:

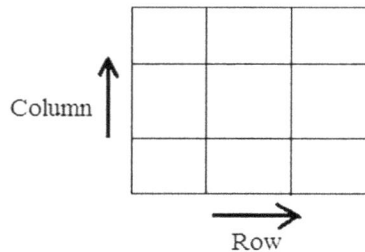

FIGURE 3.4 Representation of a 2-D array

3.8 DECLARATION OF TWO-DIMENSIONAL ARRAYS/LISTS

Just as we declared 1-D arrays, we can declare two-dimensional arrays. To declare two-dimensional arrays, we must know the name of the array, the data type of each element, and the size of each dimension (size of rows and columns).

Syntax:

```
array_name=[ [ x,x,x],
[ x,x,x],
[ x,x,x], ]
```

A two-dimensional array is also called an m × n array, as it contains m × n elements where each element in the array can be accessed by i and j, where i<=m and j<=n, and where i, j, m, n are defined as follows:

i, j = subscripts of array elements
m = number of rows
n = number of columns

For example, let us take an array/list of 3 × 3 elements. Therefore, the array/list is declared as

```
Grades=[ [x,x,x ],
[x,x,x ],
[x,x,x ] ]
```

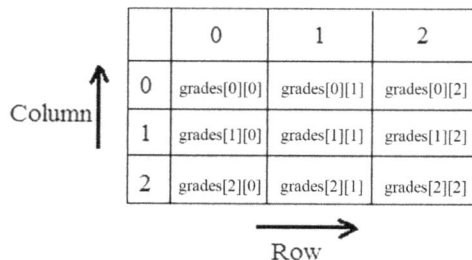

In the previous diagram, the array has 3 rows and 3 columns. The first element in the array/list is denoted by `grades [0] [0]`. Similarly, the second element is denoted by `grades [0] [1]`. The data elements in an array/list can be stored in the memory in two ways:

1. Row Major Order

In row major order, the elements of the first row are stored before the elements of the second, third, and n rows. Here, the data elements are stored on a row-by-row basis.

00	01	02	10	11	12	20	21	22					

2. Column Major Order

In column major order, the elements of the first column are stored before the elements of the second, third, and n columns. Here, the data elements are stored on a column-by-column basis.

00	10	20	01	11	21	02	12	22					

Now, we will calculate the base address of the elements in a 2-D array/list, as the computer does not store the address of each element. It just stores the address of the first element and calculates the addresses of other elements from the base address of the first element of the array/list. Hence, the addresses of other elements can be calculated from the given base address.

1. Elements in Row Major Order

$Address(A[i][j]) = Base\ Address(BA) + w(n(i - 1) + (j - 1))$

2. Elements in Column Major Order

$Address(A[i][j]) = Base\ Address(BA) + w(m(j - 1) + (i - 1))$

where w is the size in bytes needed to store one element.

Frequently Asked Questions

Q. Consider a 25 × 5 two-dimensional array/list of students that has a base address 500 and where the size of each element is 2. Calculate the address of the element *student[15][3]* assuming that the elements are stored in

    a. Row Major Order

    b. Column Major Order

Answer:

a) Row Major Order

Here, we are given that w = 2, base address = 500, n = 5, i = 15, j = 3.

Address(A[i][j]) = Base Address(BA) + w(n(i – 1) + (j – 1))

Address(student[15][3]) = 500 + 2(5(15–1) + (3 – 1))

$$= 500 + 2(5(14) + 2)$$
$$= 500 + 2(72)$$
$$= 500 + 144$$

Address(student[15][3]) = 644

b) Column Major Order

Here, we are given that w = 2, base address = 500, m = 25, i = 15, j = 3

Address(A[i][j]) = Base Address(BA) + w (m(j – 1) + (i – 1))

Address(student[15][3]) = 500 + 2(25(3 – 1) + (15 – 1))

$$= 500 + 2(25(2) + 14)$$
$$= 500 + 2(64)$$

Address(student[15][3]) = 500 + 128 = 628

3.9   OPERATIONS ON 2-D ARRAYS/LISTS

Various operations are performed on two-dimensional array/lists, which include

- **Sum** – Let A_{ij} and B_{ij} be the two matrices that are to be added together, storing the result into the third matrix C_{ij}. Two matrices are added when they are compatible with each other; that is, they should have the same number of rows and columns.

  $C_{ij} = A_{ij} + B_{ij}$

- **The Difference** – Let A_{ij} and B_{ij} be the two matrices that are to be subtracted, storing the result into a third matrix C_{ij}. The two matrices are subtracted when they are compatible with each other; that is, they should have the same number of rows and columns.

  $C_{ij} = A_{ij} - B_{ij}$

- **Product** – Let A_{ij} and B_{ij} be the two matrices that are to be multiplied together, storing the result into a third matrix C_{ij}. The two matrices are multiplied with each other if the number of columns in the first matrix is equal to the number of rows in the second matrix. Therefore, m × n matrix A can be multiplied with a p × q matrix B if n = p.

$$C_{ij} = A_{ik} \times B_{kj} \text{ for } k = 1 \text{ to } n$$

- **Transpose** – The transpose of an m × n matrix A is equal to an n × m matrix B, where

$$B_{ij} = A_{ij}.$$

// Write a program to read and display a 2 × 3 array/list.

```
# A basic code for matrix input from a user
R = int(input("Enter the number of rows:"))
C = int(input("Enter the number of columns:"))

# Initialize matrix
matrix = []
print("Enter the entries rowwise:")

# For user input
for i in range(R):          # A for loop for row entries
    a =[]
    for j in range(C):      # A for loop for column entries
        a.append(int(input()))
    matrix.append(a)

# For printing the matrix
for i in range(R):
    for j in range(C):
        print(matrix[i][j], end = " ")
    print()
```

OUTPUT

```
            Shell

     Enter the number of rows:2
     Enter the number of columns:3
     Enter the entries row wise:
     5
     7
     2
     6
     9
     0
     5 7 2
     6 9 0
     > |
```

Explanation: The above program prints a 2 × 3 matrix, i.e., a matrix that has 2 rows and 3 columns.

// Write a program to find the sum of two matrices.

```
#Program to add two matrices using a nested loop
X = [[1,2,3],
     [4 ,5,6],
     [7 ,8,9]]
Y = [[9,8,7],
     [6,5,4],
     [3,2,1]]
result = [[0,0,0],
          [0,0,0],
          [0,0,0]]
# iterate through rows
for i in range(len(X)):
# iterate through columns
    for j in range(len(X[0])):
        result[i][j] = X[i][j] + Y[i][j]

for r in result:
    print(r)
```

OUTPUT

```
    Shell

[10, 10, 10]
[10, 10, 10]
[10, 10, 10]
>
```

Explanation: The above program adds two 3 × 3 matrices, i.e., matrices with 3 rows and 3 columns.

// Write a program to find the transpose of a 3 × 3 matrix.

```
m = [[1,2],[3,4],[5,6]]
for row in m :
    print(row)
rez = [[m[j][i]

for j in range(len(m))]
```

```
for i in range(len(m[0]))]
print("\n")
for row in rez:
    print(row)
```

OUTPUT

```
                 Shell

        [1, 2]
        [3, 4]
        [5, 6]


        [1, 3, 5]
        [2, 4, 6]
        >|
```

Explanation: The above program is transposing a 2 × 3 matrix, i.e., a matrix that has 2 rows and 3 columns.

The transpose of a matrix is an operator that flips a matrix over its diagonal; that is, it switches the row and column indices of matrix A by producing another matrix.

3.10 MULTIDIMENSIONAL ARRAYS/N-DIMENSIONAL ARRAYS

A *multidimensional array* is also known as an n-dimensional array. It is an array of arrays. It has n indices in it, which justifies its name as an n-dimensional array. An n-dimensional array is an $m_1 \times m_2 \times m_3 \times \ldots \times m_n$ array, as it contains $m_1 \times m_2 \times m_3 \times \ldots \times m_n$ elements. Multidimensional arrays are declared and initialized in the same way as one-dimensional and two-dimensional arrays.

3.11 CALCULATING THE ADDRESS OF 3-D ARRAYS

Just like 2-D arrays, we can store 3-D arrays in two ways: in row major order and column major order.

1. **Elements in Row Major Order**

   Address ([i] [j] [k]) = Base Address (BA) + w (L_3 (L_2(E_1) + E_2) + E_3)

2. **Elements in Column Major Order**

   Address ([i] [j] [k]) = Base Address (BA) + w (($E_3 L_2$ + E_2) L_1 + E_1)

where L is the length of the index, L = Upper bound – Lower bound + 1, E is effective address, E = i – Lower bound.

Frequently Asked Questions

Q. Let us take a 3-D array A (4:12, –2:1, 8:14) and calculate the address of A (5, 4, 9) using row major order and column major order where the base address is 500 and w = 4.

Answer:

Length of three dimensions of A –

$L_1 = 12 – 4 + 1 = 9$

$L_2 = 1 – (–2) + 1 = 4$

$L_3 = 14 – 8 = 6$

Therefore, A contains $9 \times 4 \times 6 = 216$ *elements*

Now, $E_1 = 5 – 4 = 1$

$\qquad E_2 = 4 – (–2) = 8$

$\qquad E_3 = 9 – 8 = 1$

a. *Row Major Order*

$\quad Address(5, 4, 9 = 500 + 4(6 (4(1) + 8) + 1)$

$\qquad\qquad = 500 + 4(6 (12) + 1)$

$\qquad\qquad = 500 + 4(73)$

$\quad Address(5, 4, 9) = 500 + 292 = 792$

b. *Column Major Order*

$\quad Address(5, 4, 9) = 500 + 4((1.4 + 8)9 + 1)$

$\qquad\qquad = 500 + ((12)9 + 1)$

$\quad Address(5, 4, 9) = 500 + 145 = 645$

3.12 ARRAYS AND THEIR APPLICATIONS

Arrays are very frequently used in Python as they have various applications that are very useful. These applications include the following:

- Arrays are used for sorting the elements in ascending or descending order.
- Arrays are also used to implement various other data structures like stacks, queues, and hash tables.
- Arrays are widely used to implement matrices, vectors, and various other kinds of rectangular tables.
- Various other operations can be performed on the arrays, which include searching, merging, and sorting.

Frequently Asked Questions

Q. List some of the applications of arrays.

Answer:

1. *Arrays are very useful in storing the data in contiguous memory locations.*
2. *Arrays are used for implementing various other data structures, such as stacks and queues.*
3. *Arrays are very useful as we can perform various operations on them.*

3.13 SPARSE MATRICES

A *sparse matrix* is a matrix with a relatively high proportion of zero entries in it. A sparse matrix utilizes the memory space efficiently. The storage of null elements in the matrix is a waste of memory, so we adopt a technique to store only not-null elements in the sparse matrices.

$$\begin{bmatrix} 0 & 0 & 6 & 0 & 0 \\ 1 & 0 & 0 & 0 & 0 \\ 0 & 0 & 0 & 0 & 2 \\ 0 & 5 & 0 & 0 & 0 \end{bmatrix}$$

FIGURE 3.5 Representation of a sparse matrix

3.14 TYPES OF SPARSE MATRICES

There are three types of sparse matrices, which are

1. **Lower-Triangular Matrix –** In this type of sparse matrix, all the elements above the main diagonal must have a zero value, or in other words, we can say that all the elements below the main diagonal should contain non-zero elements only. This type of matrix is called a *lower triangular matrix*.

$$\begin{bmatrix} 5 & 0 & 0 & 0 \\ 4 & 6 & 0 & 0 \\ -3 & 9 & -5 & 0 \\ 2 & 1 & 7 & 3 \end{bmatrix}$$

FIGURE 3.6 Lower-triangular matrix

2. **Upper-Triangular Matrix –** In this type of sparse matrix, all the elements above the main diagonal should contain non-zero elements only, or in other words, we can say that all the elements below the main diagonal should have a zero value. This type of matrix is called an *upper-triangular matrix*.

$$\begin{bmatrix} 1 & 2 & 3 & 4 \\ 0 & 6 & -1 & 5 \\ 0 & 0 & -7 & 8 \\ 0 & 0 & 0 & 9 \end{bmatrix}$$

FIGURE 3.7 Upper-triangular matrix

3. **Tri-diagonal Matrix –** In this type, elements with a non-zero value can appear only on the diagonal or adjacent to the diagonal. This type of matrix is a *tri-diagonal matrix*.

$$\begin{bmatrix} 6 & 2 & 0 & 0 \\ 8 & 9 & -2 & 0 \\ 0 & 5 & -7 & 3 \\ 0 & 0 & 1 & 4 \end{bmatrix}$$

FIGURE 3.8 Tri-diagonal matrix

3.15 REPRESENTATION OF SPARSE MATRICES

There are two ways in which the sparse matrices can be represented, which are

1. **Array/list Representation/3-Tuple Representation** – This representation contains three rows in which the first row represents the number of rows, columns, and non-zero entries/values in the sparse matrix. Elements in the other rows give information about the location and value of non-zero elements.

   For example, let us consider a sparse matrix.

$$\begin{bmatrix} 0 & 0 & 0 & 0 & 1 \\ 0 & 0 & 0 & 0 & 0 \\ 0 & 0 & 3 & 0 & 0 \\ 0 & 5 & 0 & 0 & 0 \end{bmatrix}$$

   An array representation of the previous sparse matrix is shown in the table.

Row	Column	Non-Zero Value
0	4	1
2	2	3
3	1	5

2. **Linked Representation** – A sparse matrix can also be represented in a linked way. In this representation, we store the number of rows, columns, and non-zero entries in a single node, and there is an address field that stores the next location. Let us consider the following sparse matrix.

$$\begin{bmatrix} 0 & 6 & 0 & 0 & 0 \\ 0 & 0 & 0 & 0 & 0 \\ 0 & 0 & 0 & 8 & 3 \end{bmatrix}$$

The linked representation of the previous sparse matrix is as follows.

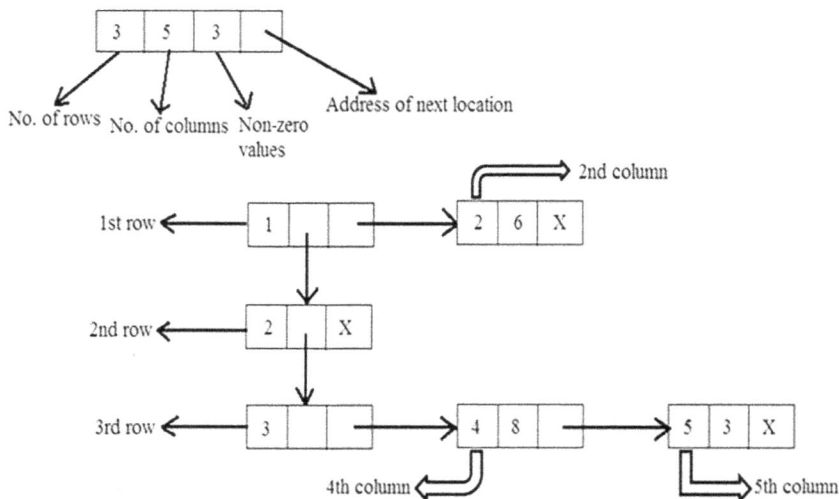

FIGURE 3.9 Linked representation of a sparse matrix

Frequently Asked Questions

Q. Explain the sparse matrix.

Answer:

A matrix in which the number of zero entries is much higher than the number of non-zero entries is called a sparse matrix. The natural method of representing matrices in memory as two-dimensional arrays may not be suitable for sparse matrices. One may save space by storing only non-zero entries. We can represent a sparse matrix by using a three-tuple method of storage:

1. *Row Major Method*
2. *Column Major Method*

3.16 SUMMARY

- An array is a collection of homogeneous (similar) types of data elements in contiguous memory. An array is a linear data structure because all elements of an array are stored in linear order.

- A list is a data structure in Python that is a mutable, or changeable, ordered sequence of elements. Each element or value that is inside of a list is called an item. Just as strings are defined as characters between quotes, lists are defined by having values between square brackets [ ].

- The initialization of the elements of an array/list at compile time is done in the same way as when we initialize the normal or ordinary variables at the time of their declaration.

- The initialization of the elements of an array/list at runtime refers to the method of inputting the values from the keyboard.

- The address of the elements in a 1-D array/list can be calculated very easily, as an array/list stores all its data elements in contiguous memory locations, storing the base address.

- Traversing an array/list means to access each and every element in an array/list exactly once so that it can be processed.

- The insertion of an element in an array/list refers to the operation of adding an element to the array/list. It can be done in two ways.

- Deleting an element from an array/list refers to the operation of the removal of an element from an array/list. Deletion is also done in two ways.

- Searching for an element in an array/list means finding whether a particular value exists in an array/list or not. If that particular value is found, then the search is said to be successful and the position/location of that particular value is returned. If the value is not found, then searching is said to be unsuccessful.

- The merging of two array/lists means copying the elements of the first and second array/lists into a third array/list.

- Sorting an array/list means arranging the data elements of a data structure in a specified order in either ascending or descending order.

- Unlike 1-D arrays, 2-D arrays are organized in the form of grids or tables. They are collections of 1-D arrays.

- A multidimensional array is also known as an n-dimensional array. It is an array of arrays. It has n indices in it, which also justifies its name as an n-dimensional array.

- A sparse matrix is a matrix with a relatively high proportion of zero entries in it. A sparse matrix is used because it utilizes the memory space efficiently.

3.17 EXERCISES

3.17.1 Theory Questions

Q1. What is meant by an array and how is it represented in the memory?

Q2. What is a list?

Q3. What are the differences between a list and an array?

Q4. What are the various operations that can be performed on arrays/lists? Discuss them in detail.

Q5. Explain the concept of two-dimensional arrays.

Q6. In how many ways can arrays/lists be initialized? Explain in detail.

Q7. What is meant by sorting an array/list? Explain.

Q8. Explain the process of merging two arrays/lists along with the algorithm.

Q9. Give some of the applications of arrays.

Q10. What is a sparse matrix? Explain its types.

Q11. Consider a three-dimensional array A (2:6, -1:7, 9:10) and calculate the address of A (9, 6, 8) using row major order and column major order when the base address is 2000 and w = 4.

Q12. Explain the linked representation of sparse matrices in detail.

Q13. Write the formulae for calculating the addresses of elements in row major and column major order in 2-D and 3-D arrays.

3.17.2 Programming Questions

Q1. Write a Python program to traverse an entire array/list.

Q2. Write a Python program to perform an insertion at a specified position in a one-dimensional array/list.

Q3. Write a Python program to multiply two matrices.

Q4. Write a Python program which reads a matrix and displays the

    a. Sum of its rows' elements

    b. Sum of its columns' elements

    c. Sum of its diagonal's elements

Q5. Write a Python program to perform the deletion of an element.

Q6. Write a menu-driven Python program to perform various insertions and deletions in an array/list using the switch case.

Q7. Write an algorithm for reversing an array/list.

Q8. Write a program that reads an array/list of 50 integers. Display all the pairs of elements whose sum is 25.

Q9. Write a Python program to read an array/list of 10 integers and then find the smallest and largest numbers in the array/list.

Q10. Write a Python program to add two sparse matrices.

3.17.3 Multiple Choice Questions

Q1. The elements of an array/list are always stored in _____ memory locations.

a. Random

b. Sequential

c. Both

d. None of these

Q2. Array [5] = 19 initializes the _____ element of the array with value 19.

a. 4th

b. 5th

c. 6th

d. 7th

Q3. By default, the first subscript of the array/list is _____.

a. 2

b. 1

c. −1

d. 0

Q4. A multidimensional array, in simple terms, is an

   a. array of arrays

   b. array of addresses

   c. Both

   d. None of the above

Q5. What is the output when we execute `list("hello")`?

   a. ['h', 'e', 'l', 'l', 'o']

   b. ['hello']

   c. ['llo']

   d. ['olleh']

Q6. A loop is used to access all the elements of an array/list.

   a. False

   b. True

   c. None of the above

Q7. Which of the following commands will create a list?

   a. `list1 = list()`

   b. list1 = []

   c. list1 = list([1, 2, 3])

   d. all of the above

Q8. A sparse matrix has a _____.

   a. A high proportion of zeroes

   b. A low proportion of zeroes

   c. Both (a) and (b)

   d. None of the above

## CHAPTER 4

LINKED LISTS

4.1 INTRODUCTION

We learned that an array is a collection of data elements stored in contiguous memory locations. We also studied that arrays were static; that is, the size of the array must be specified when declaring the array, which limits the number of elements to be stored in the array. For example, if we have an array declared as int[]array = new int[15], then the array can contain a maximum of 15 elements and not more than that. This method of allocating memory is good when the exact number of elements is known. However, if we are not sure of the number of elements, there will be a problem, because the data structures we use to make the program efficient should consume little memory space and a minimal amount of time. To overcome this problem, we use linked lists.

4.2 DEFINITION OF A LINKED LIST

A *linked list* is a linear collection of data elements. These data elements are called *nodes*, and they point to the next node. A linked list is a data structure that can be used to implement other data structures such as stacks, queues, and trees. A linked list is a sequence of nodes in which each node contains one or more data fields that point to the next node. Also, linked lists are dynamic; that is, memory is allocated when required. There is no need to know the exact size or the exact number of elements as in the case of arrays. Figure 4.1 contains an example of a simple linked list that contains five nodes.

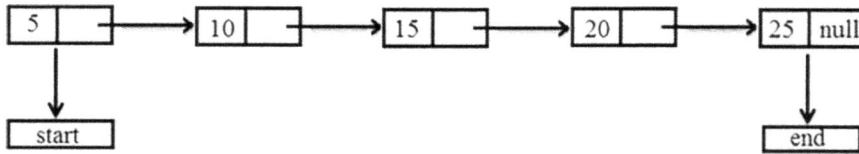

FIGURE 4.1 A linked list

In Figure 4.1, we have a linked list in which each node is divided into two parts:

1. The first part contains information/data.

2. The second part contains the address of the next node.

The last node will not have any next node connected to it, so it will store a special value called NULL. Usually, NULL is defined by –1. Therefore, the NULL node represents the end of the linked list. There is another special node, START, that stores the address of the first node of the linked list. Therefore, the START node represents the beginning of the linked list. If START = NULL, then the linked list is empty. A linked list is known as a self-referential data type or a self-referential structure because each node points to another node that is of the same type.

The self-referential structure in a linked list is as follows:

```
class ListNode :
      def __init__ ( self, data ) :
           self.data = data
           self.next = None
```

Practical Application:

- A simple real-life example is how each car on a train is connected to its previous and next car (except the first and last). In terms of programming, consider the car body as a node and the connectors as links to the previous and next nodes.

- The brain is also a good example of a linked list. In the initial stages of learning something by heart, the natural process is to link one item to another item. It's a subconscious act. Also, when we forget something and try to remember it, our brain follows associations and tries to link one memory with another until we finally recall the lost memory.

Frequently Asked Questions

Q1. Define the linked list.

Answer:

A linked list is a linear collection of data elements, called nodes, where the linear order is given using nodes. It is a dynamic data structure. For every data item in a linked list, there is an associated node that gives the memory location of the next data item in the linked list. The data items in the linked list are not in consecutive memory locations.

Frequently Asked Questions

Q2. List the advantages and disadvantages of a linked list.

Answer:

Advantages of linked lists

1. *Linked lists are dynamic data structures; that is, they can grow or shrink during the execution of the program.*

2. *Linked lists have efficient memory utilization. Memory is allocated whenever it is required, and it is de-allocated whenever it is no longer needed.*

3. *Insertion and deletion are easier and efficient.*

4. *Many complex applications can be easily carried out with linked lists.*

Disadvantages of linked lists

1. *They consume more space because every node requires an additional node to store the address of the next node.*

2. *Searching a particular element in the list is difficult and time-consuming.*

4.3 MEMORY ALLOCATION IN A LINKED LIST

The process or concept of linked lists supports dynamic memory allocation. *Dynamic memory allocation* is the process of allocating memory during the execution of the program or the process of allocating memory to the variables at runtime. Until now, we have studied arrays in which we declared the size of the array initially, such as array[50]. This statement after execution allocates the memory for 50 integers. But there can be a problem if we use only 30% of the memory and the rest of the allocated memory is wasted. Therefore, to overcome this problem of wasted memory space (or, in other words, to utilize the memory efficiently), dynamic memory allocation is used, which allows us to allocate/reserve the memory that is required. Hence, we overcome the problem of wasted memory space as in the case of arrays. Dynamic memory allocation is best when we are not aware of the memory requirements in advance.

4.4   TYPES OF LINKED LISTS

Different types of linked lists are discussed in this section. These include the following:

1.   Singly Linked List

2.   Circular Linked List

3.   Doubly Linked List

4.   Header Linked List

4.4.1 Singly Linked List

A *singly linked list* is the simplest type of linked list in which each node contains some information/data and only one node, which points to the next node in the linked list. The traversal of data elements in a singly linked list can be done only in one way.

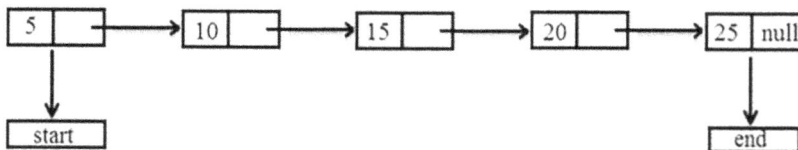

FIGURE 4.2 Singly-linked list

4.4.2 Operations on a Singly Linked List

Various operations can be performed on a singly linked list, which includes
- Traversing a linked list
- Searching for a given value in a linked list
- Inserting a new node in a linked list

- Deleting a node from a linked list
- Concatenation of two linked lists
- Sorting a linked list
- Reversing a linked list

a. Traversing a linked list

Traversing a linked list means accessing all the nodes of the linked list exactly once. A linked list will always contain a START node, which stores the address of the first node of the linked list and which also represents the beginning of the linked list, and a NULL node which represents the end of the linked list. For traversing a linked list, we use another node variable, NODE, which points to the node that is currently being accessed. The algorithm for traversing a linked list is shown as follows.

Algorithm for traversing a linked list

```
Step 1: Set NODE = START
Step 2: Repeat Steps 3 & 4 while NODE != NULL
Step 3: Print NODE. INFO
Step 4: Set NODE = NODE. NEXT
[End of Loop]
Step 5: Exit
```

b. Searching for a given value in a linked list

Searching for a value in a linked list means to find a particular element/value in the linked list. As we discussed earlier, a node in a linked list contains two parts: one part is the information part and the other is the address part. Hence, searching refers to the process of finding whether the given value exists in the information part of any node. If the value is present, then the address of that particular value is returned and the search is said to be successful; otherwise, the search is unsuccessful. A linked list will always contain a START node that stores the address of the first node of the linked list and represents the beginning of the linked list and a NULL node that represents the end of the linked list. There is another variable, NODE, that points to the current node being accessed. SEARCH_VAL is the value to be searched in the linked list, and POS is the position/address of the node at which the value is found. The algorithm for searching a value in a linked list is given as follows:

Algorithm to search a value in a linked list

```
Step 1: Set NODE = START
Step 2: Repeat Step 3 while NODE != NULL
Step 3: IF SEARCH_VAL = NODE. INFO
                Set POS = NODE
                Print Successful Search!!
                Go to Step 5
[End of If]
ELSE
                Set NODE = NODE. NEXT
[End of Loop]
Step 4: Print Unsuccessful Search!!
Step 5: Exit
```

For example, if we have a linked list and we are searching for 15 in the list, then the steps are as shown in Figure 4.3.

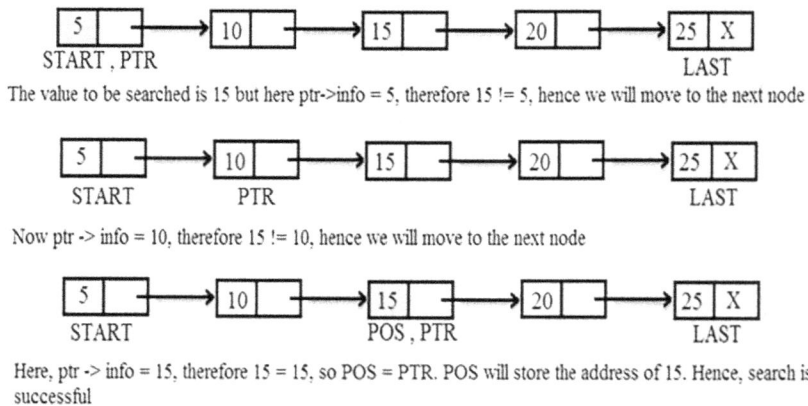

The value to be searched is 15 but here ptr->info = 5, therefore 15 != 5, hence we will move to the next node

Now ptr -> info = 10, therefore 15 != 10, hence we will move to the next node

Here, ptr -> info = 15, therefore 15 = 15, so POS = PTR. POS will store the address of 15. Hence, search is successful

FIGURE 4.3 An example of searching a linked list

c. Inserting a new node in a linked list

Here, we discuss how a new node is inserted in an existing linked list. The three cases in the insertion process include the following

1. A new node is inserted at the beginning of the linked list.

2. A new node is inserted at the end of the linked list.

3. A new node is inserted after the given node in a linked list.

1. Inserting a new node at the beginning of a linked list

In the case of inserting a new node at the beginning of a linked list, we first check the overflow condition, which is whether the memory is available for a new node. If the memory is not available, then an overflow message is displayed; otherwise, the memory is allocated for the new node. Now, we initialize the node with its info part, and its address part contains the address of the first node of the list, which is the START node. Hence, the new node is added as the first node in the list and the START node will point to the first node of the list. Now to understand better, let us take an example. Consider the linked list with five nodes shown in Figure 4.4; a new node will be inserted at the beginning of the linked list.

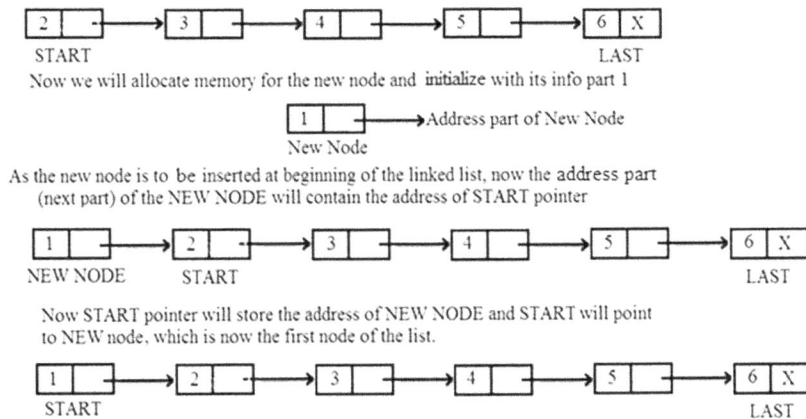

FIGURE 4.4 Inserting a new node at the beginning of a linked list

From the previous example, it is clear how a new node is inserted in an already existing linked list. Let us now examine its algorithm.

Algorithm for inserting a new node at the beginning of a linked list

```
Step 1: START
Step 2: IF NODE = NULL
            Print OVERFLOW
            Go to Step 8
[End of If]
Step 3: Set NEW NODE = NODE
Step 4: Set NODE = NODE. NEXT
Step 5: Set NEW NODE. INFO = VALUE
```

```
Step 6: Set NEW NODE. NEXT = START
Step 7: Set START = NEW NODE
Step 8: EXIT
```

2. Inserting a new node at the end of a linked list

To insert the new node at the end of the linked list, we first check the overflow condition, which is whether the memory is available for a new node. If the memory is not available, then an overflow message is displayed; otherwise, the memory is allocated for the new node. Then a NODE variable is made, which initially points to START and is used to traverse the linked list until it reaches the last node. When it reaches the last node, the NEXT part of the last node stores the address of the new node, and the NEXT part of the NEW NODE contains NULL, which denotes the end of the linked list. Let us understand this with the help of an algorithm.

Algorithm for inserting a new node at the end of a linked list

```
Step 1: START
Step 2: IF NODE = NULL
            Print OVERFLOW
            Go to Step 10
[End of If]
Step 3: Set NEW NODE = NODE
Step 4: Set NODE = NODE. NEXT
Step 5: Set NEW NODE. INFO = VALUE
Step 6: Set NEW NODE. NEXT = NULL
Step 7: Set NODE = START
Step 8: Repeat Step 8 while NODE. NEXT != NULL
            Set NODE = NODE. NEXT
[End of Loop]
Step 9: Set NODE. NEXT = NEW NODE
Step 10: EXIT
```

From the previous algorithm, we understand how to insert a new node at the end of the already existing linked list. Now let's consider the following example. Consider the linked list with four nodes shown in Figure 4.5; a new node is inserted at the end of the linked list.

Now we will allocate memory for the new node and initialize with its info part

As the NEW NODE is to be inserted at the end of the linked list, so we will use another pointer variable PTR which will initially point to the START.

Now in the next step PTR variable will be moved till the last node until it has found a NULL. And finally it reaches the last node.

After reaching to the last node now we will store the address of the NEW NODE in the address part(next part) of PTR so that it starts pointing to the NEW NODE.

Finally, NEW NODE is inserted at the end of the linked list.

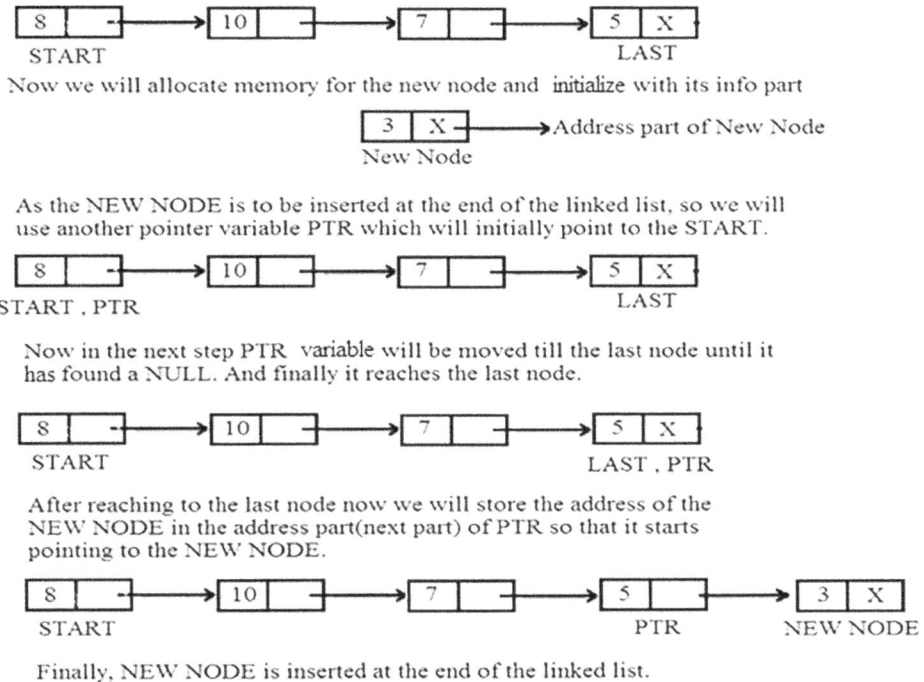

FIGURE 4.5 Inserting a new node at the end of a linked list

3. Inserting a new node after a node in a linked list

In this case, a new node is inserted after a given node in a linked list. As in the other cases, we again check the overflow condition. If memory for the new node is available, it will be allocated; otherwise, an overflow message is printed. Then a NODE variable is made that initially points to START, and the NODE variable is used to traverse the linked list until it reaches the value/node, after which the new node is inserted. When it reaches that node/ value, then the NEXT part of that node stores the address of the new node and the NEXT part of the NEW NODE stores the address of its next node in the linked list. Let us understand this with the help of an example. Consider a linked list with four nodes, and a new node is to be inserted after the given node, as shown in Figure 4.6.

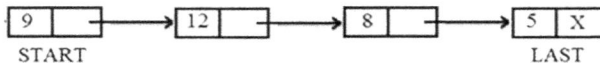

START   LAST

Now we will allocate memory for the new node and initialize with its info part 1

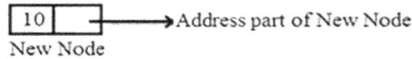

Two variables PTR and PREV are used which initially point to the START such that PTR, PREV, START all point to the first node.

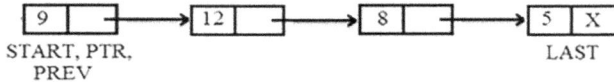

Now, PTR and PREV are moved till info part of PREV is equal to value of node after which NEW NODE is to be inserted. PREV will always point to the node before PTR.

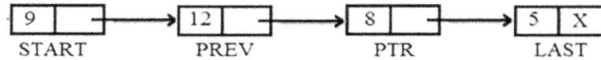

Hence, NEW NODE is to be inserted after 12. Also address part of PREV will now store the address of NEW NODE. Similarly, address part of NEW NODE will store the address of PTR.

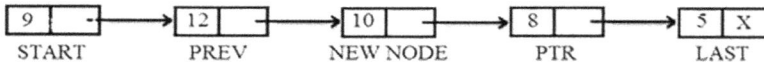

FIGURE 4.6 Inserting a new node after a given node in a linked list

From the previous example, we learned how a node can be inserted after a given node. Now let's take a look at the algorithm.

Algorithm for inserting a new node after a given node in a linked list

```
Step 1: START
Step 2: IF NODE = NULL
             Print OVERFLOW
             Go to Step 10
[End of If]
Step 3: Set NEW NODE = NODE
Step 4: Set NODE = NODE. NEXT
Step 5: Set NEW NODE. INFO = VALUE
Step 6: Set NODE = START
Step 7: Set PREV = NODE
Step 8: Repeat Step 8 while PREV. INFO != GIVEN_VAL
             Set PREV = NODE
             Set NODE = NODE. NEXT
[End of Loop]
Step 9: Set PREV. NEXT = NEW NODE
Step 10: Set NEW NODE. NEXT = NODE
Step 11: EXIT
```

d. Deleting a node from a linked list

In this section, we learn how a node is deleted from an already existing linked list. We discuss three cases in the deletion process which include

1. A node is deleted from the beginning of the linked list.

2. A node is deleted from the end of the linked list.

3. A node is deleted after a given node from the linked list.

1. Deleting a node from the beginning of the linked list

To delete a node from the beginning of a linked list, we first check the underflow condition, which occurs when we try to delete a node from a linked list that is empty. This situation exists when the START node is equal to NULL. If the condition is true, then the underflow message is printed on the screen; otherwise, the node is deleted from the linked list. Consider the linked list with five nodes in Figure 4.7; the node is deleted from the beginning of the linked list.

From the below example, it is clear how a node is deleted from an already existing linked list. Let us now take a look at its algorithm.

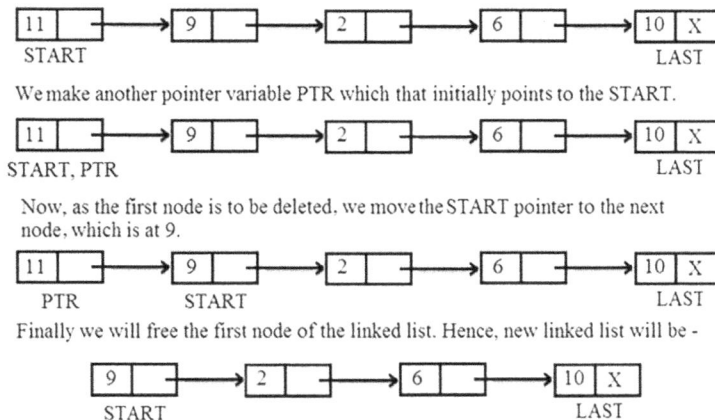

FIGURE 4.7 Deleting a node from the beginning of a linked list

Algorithm for deleting a node from the beginning of a linked list

```
Step 1: START
Step 2: IF START = NULL
            Print UNDERFLOW
[End Of If]
Step 3: Set NODE = START
Step 4: Set START = START. NEXT
Step 5: FREE NODE
Step 6: EXIT
```

In the previous algorithm, we check for the underflow condition, that is, whether there are any nodes present in the linked list. If there are no nodes, then an underflow message is printed; otherwise, we move to Step 3, where we are initializing NODE to START, that is, NODE now stores the address of the first node. In the next step, START is moved to the second node, as now START stores the address of the second node. Hence, the first node is deleted and the memory, which was occupied by NODE (initially the first node of the list), is free.

2. Deleting a node from the end of the linked list

To delete a node from the end of the linked list, we first check the under-flow condition. This situation exists when the START node is equal to NULL. Hence, if the condition is true, then the underflow message is printed on the screen; otherwise, the node is deleted from the linked list. Consider the linked list with five nodes shown in Figure 4.8; the node is deleted from the end of the linked list.

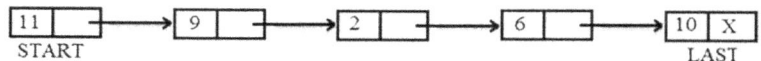

We will make another pointer variable PTR which will initially point to the START.

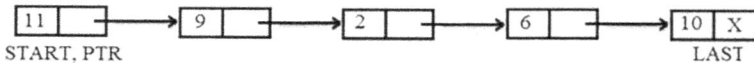

We will make another pointer variable PREV which will always point to the node just before the node pointed by PTR. Move PTR till the next part of the PTR = NULL.

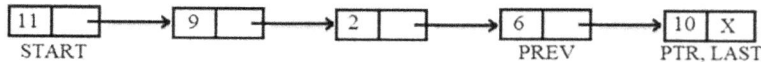

Finally set the next part of PREV is equal to NULL. Also LAST will now point to PREV and the last node is deleted.

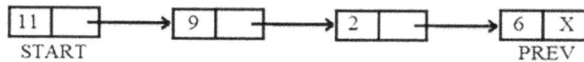

FIGURE 4.8 Deleting a node from the end of a linked list

Let us now look at the algorithm of deleting a node from the end of a linked list.

Algorithm for deleting a node from the end of a linked list

```
Step 1: START
Step 2: IF START = NULL
              Print UNDERFLOW
[End Of If]
```

```
Step 3: Set NODE = START
Step 4: Repeat while NODE. NEXT != NULL
              Set PREV = NODE
              Set NODE = NODE. NEXT
[End of Loop]
Step 5: Set PREV. NEXT = NULL
Step 6: FREE NODE
Step 7: EXIT
```

In the algorithm, we again check for the underflow condition. If the condition is true, then the underflow message is printed; otherwise, NODE is initialized to the START node, that is, the NODE is pointing to the first node of the list. In the loop, we have taken another node variable PREV, which will always point to one node before the NODE node. After reaching the last node of the list, we set the next part of PREV to NULL. Therefore, the last node is deleted, and the memory that was occupied by the NODE node is now free.

3. Deleting a node after a given node from the linked list

In the case of deleting a node after a given node from the linked list, we again check the underflow condition as we checked in both the other cases. This situation exists when the START node is equal to NULL. Hence, if the condition is true, then the underflow message is printed; otherwise, the node is deleted from the linked list. Consider the linked list with five nodes shown in Figure 4.9; the node will be deleted after a given node from the linked list.

Now let us examine the previous case with the help of an algorithm.

Algorithm for deleting a node after a given node from the linked list

```
Step 1: START
Step 2: IF START = NULL
              Print UNDERFLOW
[End Of If]
Step 3: Set NODE = START
Step 4: Set PREV = START
Step 5: Repeat while PREV. INFO != GIVEN_VAL
              Set PREV = NODE
              Set NODE = NODE. NEXT
[End of Loop]
Step 6: Set PREV. NEXT = NODE. NEXT
Step 7: FREE NODE
Step 8: EXIT
```

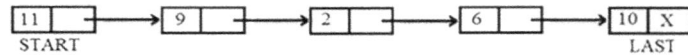

We will make another pointer variable PTR which will initially point to the START.

Two other pointer variables PTR and PREV are used which will initially point to START such that PTR, PREV, START all point to the first node.

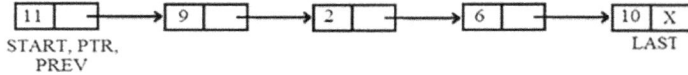

Now PTR and PREV are moved such that PREV points to the node containing the value after which the node is to be deleted and PTR points to the node which is to be deleted.

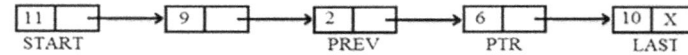

Now we know that node containing 6 is to be deleted from the linked list. So, we set the NEXT part of PREV to the NEXT part of PTR such that PREV will now point to LAST node.

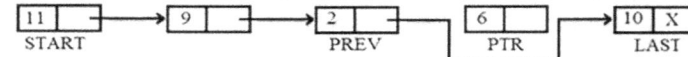

Now PTR is deleted and new list after deletion is given below -

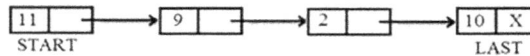

FIGURE 4.9 Deleting a node after a given node from the linked list

In the previous algorithm, we first check for the underflow condition. If the condition is true, then the underflow message is printed; otherwise, NODE is initialized to the START node, that is, the NODE is pointing to the first node of the list. In the loop, we have taken another node variable PREV, which always points one node before the NODE node. After reaching the node containing the given value which is to be deleted, we set the next node of the node containing the given value to the address contained in the next part of the succeeding node. Therefore, the node is deleted and the memory that was being occupied by NODE is now free.

e. Concatenation of two linked lists

A concatenated linked list is created by the process of concatenating two different-sized linked lists into one linked list. Let us consider concatenation with the help of a function.

```
def list_concat(A, B):
  while A.next != None:
A = A.next
A.next = B
  return A
```

f. Sorting a linked list

Sorting is the process of arranging the data elements in a sequence, either in ascending order or in descending order. In this, we are arranging the information on the linked list in a sequence.

g. Reversing a linked list

In the process of reversing a linear linked list, we will take three node variables, that is, PREV, NODE, and NEW, which hold the addresses of the previous node, current node, and the next node, respectively, in the linked list. We begin with the address of the first node, which is held in another node variable START, which is assigned to NODE, and PREV is assigned to NULL.

```python
def reverse(self):
        prev = None
        current = self.head
        while(current is not None):
            next = current.next
  current.next = prev
            prev = current
            current = next
        self.head = prev
```

Here is a program to implement a singly linked list.

```python
# Node class
class Node:
    # Function to initialize the node object
    def __init__(self, data):
        self.data = data # Assign data
        self.next = None # Initialize next as null
# Linked List class contains a Node object
class LinkedList:
    # Function to initialize head
    def __init__(self):
        self.head = None
    # insertion method for the linked list
    def insert(self, data):
        newNode = Node(data)
        if(self.head):
            current = self.head
            while(current.next):
                current = current.next
            current.next = newNode
        else:
            self.head = newNode
    # print method for the linked list
    def printLL(self):
        current = self.head
        while(current):
            print(current.data)
            current = current.next
```

```python
        # delete the first occurence of key in linked list
    def deleteNode(self,key):
        # Store head node
        temp = self.head
        # If head node itself holds the key to be deleted
        if (temp is not None):
            if (temp.data == key):
            self.head = temp.next
            temp = None
            return
      # Search for the key to be deleted, keep track of the
      # previous node as we need to change 'prev.next'
      while(temp is not None):
            if temp.data == key:
                break
            prev = temp
            temp = temp.next

     # if key was not present in linked list
     if(temp == None):
         return
     # Unlink the node from linked list
     prev.next = temp.next
     temp = None
```

OUTPUT

```
>>> sll=LinkedList()
>>> sll.insert(7)
>>> sll.insert(5)
>>> sll.printLL()
7
5

>>> sll.insert(9)
>>> sll.insert(3)
>>> sll.printLL()
7
5
9
3
```

```
>>> sll.deleteNode(5)
>>> sll.printLL()
7
9
3
>>> |
```

Explanation: The above program has the insert, deleteNode, and printLL functions for the singly linked list.

- The insert function add nodes at the beginning of the list.
- The deleteNode function removes a given node from the list.
- The printLL function prints all the nodes of the list.

Let us now consider another type of linked list: the circular linked list.

4.4.3 Circular Linked Lists

Circular linked lists are a type of singly linked list in which the address part of the last node stores the address of the first node, unlike in singly linked lists in which the address part of the last node stores a unique value, NULL. While traversing a circular linked list, we can begin from any node and traverse the list in any direction because a circular linked list does not have a first or last node. The memory declarations for representing a circular linked list are the same as for a linear linked list.

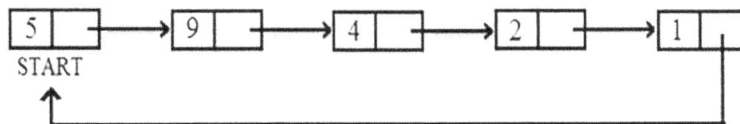

FIGURE 4.10 Circular linked list

4.4.4 Operations on a Circular Linked List

Various operations can be performed on a circular linked list, which include

a. Inserting a new node in a circular linked list

b. Deleting a node from a circular linked list

Let us now discuss both these cases in detail.

a. Inserting a new node in a circular linked list

Here, we learn how a new node is inserted in an existing linked list. We discuss the cases in the insertion process which include when

1. A new node is inserted at the beginning of the circular linked list.

2. A new node is inserted at the end of the circular linked list.

3. A new node is inserted after a given node (the same as that for a singly linked list).

1. Inserting a new node at the beginning of a circular linked list

In the case of inserting a new node at the beginning of a circular linked list, we first check the overflow condition, that is, whether the memory is available for a new node. If the memory is not available, then an overflow message is printed; otherwise, the memory is allocated for the new node. Then we initialize the node with its info part, and its address part contains the address of the first node of the list, which is the START node. Hence, the new node is added as the first node in the list, and the START node points to the first node of the list. Now let us take an example. Consider a linked list with four nodes as shown in Figure 4.11; a new node is inserted at the beginning of the circular linked list.

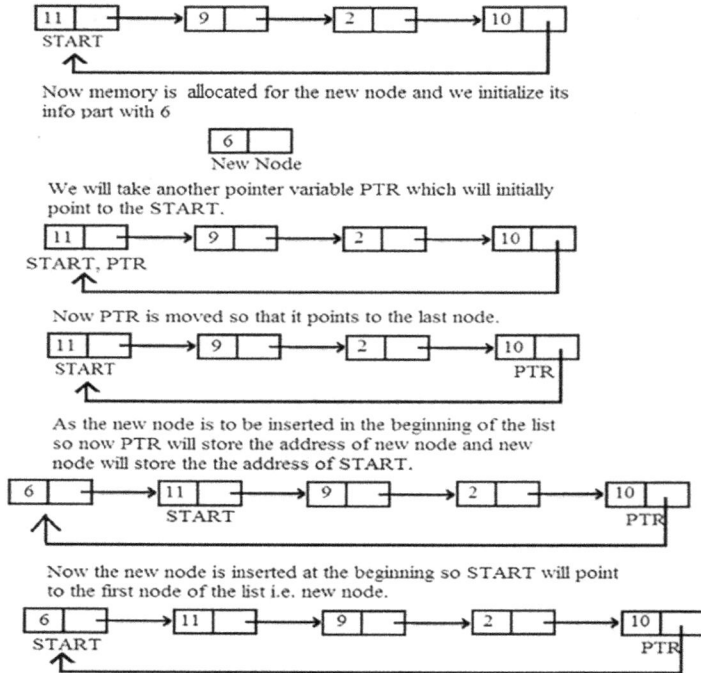

FIGURE 4.11 Inserting a new node at the beginning of a circular linked list

Now let us examine the previous case with the help of an algorithm.

Algorithm for inserting a new node at the beginning of a circular linked list

```
Step 1: START
Step 2: IF TEMP = NULL
            Print OVERFLOW
        [End Of If]
Step 3: Set NEW NODE = TEMP
Step 4: Set NEW NODE. INFO = VAL
Step 5: Set NEW NODE. NEXT = START
Step 6: Set END. NEXT = NEW NODE
Step 7: Set START = NEW NODE
Step 8: EXIT
```

2. Inserting a new node at the end of a circular linked list

In this case, we first check the overflow condition, that is, whether the memory is available for a new node. If the memory is not available, then an overflow message is printed; otherwise, the memory is allocated for the new node. Then a NODE variable is made, which initially points to START, and the NODE variable is used to traverse the linked list until it reaches the last node. When it reaches the last node, the NEXT part of the last node stores the address of the new node and the NEXT part of the NEW NODE contains the address of the first node of the linked list, which is denoted by START. Let us understand it with the help of an algorithm.

Algorithm for inserting a new node at the end of a circular linked list

```
Step 1: START
Step 2: IF TEMP = NULL
            Print OVERFLOW
[End Of If]
Step 3: Set NEW NODE = TEMP
Step 4: Set NEW NODE. INFO = VAL
Step 5: Set NEW NODE. NEXT = START
Step 6: Set END. NEXT = NEW NODE
Step 7: Set END = NEW NODE
Step 8: EXIT
```

Let us take an example. Consider a linked list with four nodes, as shown in Figure 4.12; a new node is inserted at the end of the circular linked list.

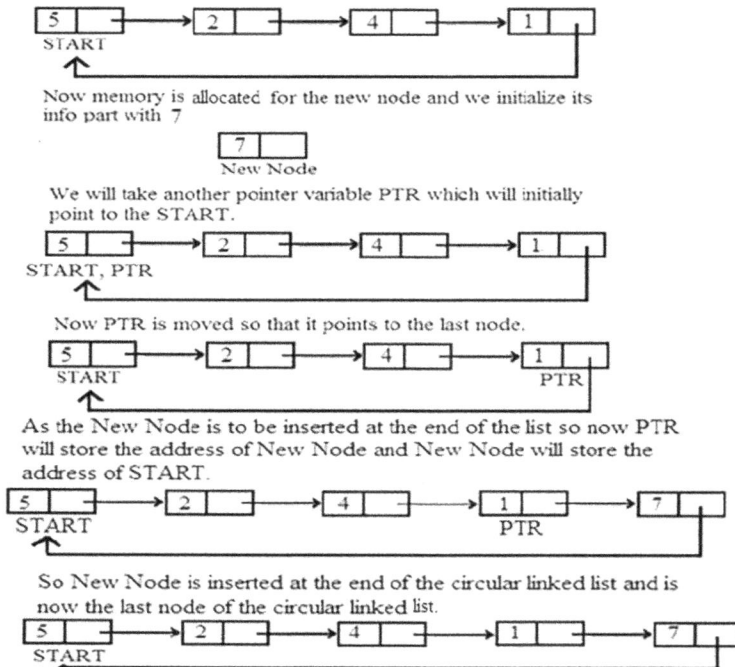

FIGURE 4.12 Inserting a new node at the end of a circular linked list

b. Deleting a node from a circular linked list

In this section, we learn how a node is deleted from an already existing circular linked list. We discuss several cases in the deletion process, including when

1. A node is deleted from the beginning of the circular linked list.

2. A node is deleted from the end of the circular linked list.

3. A node is deleted after a given node (same as that for a singly linked list).

1. Deleting a node from the beginning of a circular linked list

In the case of deleting a node from the beginning of a linked list, we first check the underflow condition, which occurs when we try to delete a node from the linked list that is empty. This situation exists when the START node is equal to NULL. Hence, if the condition is true, then an underflow message is displayed; otherwise, the node is deleted from the linked list. Consider a linked list with four nodes, as shown in Figure 4.13; the first node is deleted from the linked list.

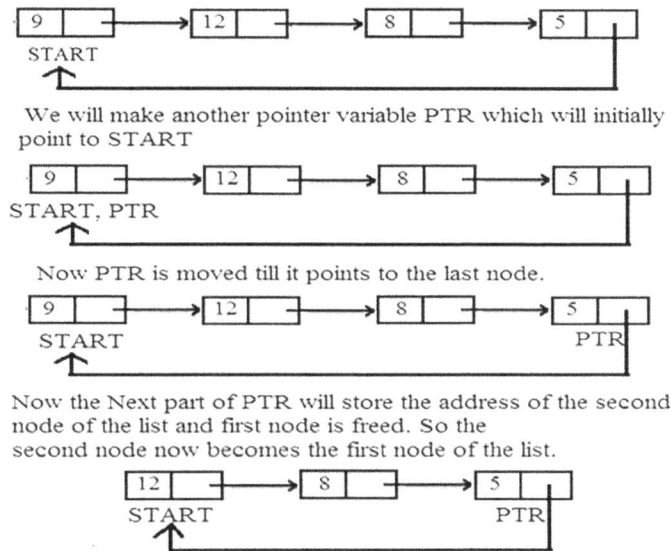

FIGURE 4.13 Deleting a node from the beginning of a circular linked list

From the previous example, it is clear how a node is deleted from an already existing linked list. Let us now examine its algorithm.

Algorithm for deleting a node from the beginning of a circular linked list

```
Step 1: START
Step 2: IF START = NULL
            Print UNDERFLOW
        [End Of If]
Step 3: Set END.NEXT = START.NEXT
Step 4: Set START = START.NEXT
Step 4: EXIT
```

The previous algorithm shows how a node is deleted from the beginning of the linked list. First, we check with the underflow condition. Now a node variable NODE is used which traverses the entire list until it reaches the last node of the list. We change the next part of the NODE to store the address of the second node of the list. Hence, the memory that occupied the first node is freed. Finally, the second node now becomes the first node of the linked list.

2. Deleting a node from the end of a circular linked list

In this case, we first check the underflow condition, which is when we try to delete a node from the linked list that is empty. This situation occurs when the START node is equal to NULL. Hence, if the condition is true, then an underflow message is printed; otherwise, the node is deleted from the linked list. Consider a linked list with four nodes as shown in Figure 4.14; the last node is deleted from the linked list.

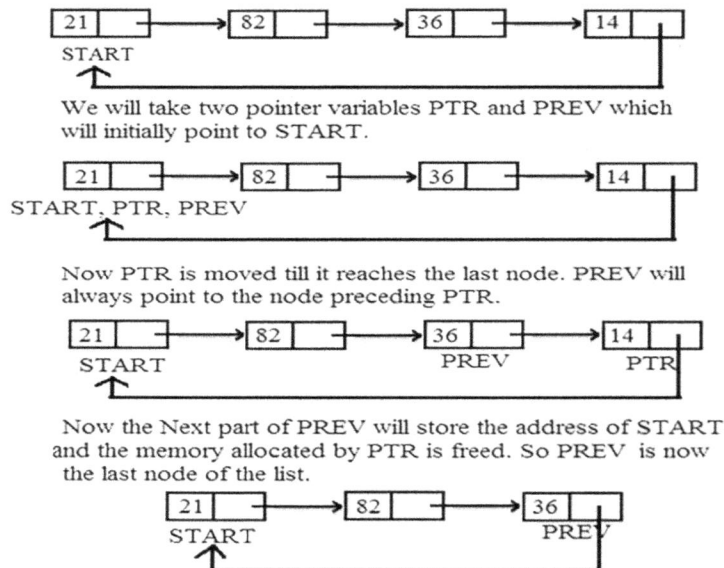

FIGURE 4.14 Deleting a node from the end of a circular linked list

Let us now examine its algorithm.

Algorithm for deleting a node from the end of a circular linked list

```
Step 1: START
Step 2: IF START = NULL
            Print UNDERFLOW
        [End Of If]
Step 3: Set NODE = START
Step 4: Repeat while NODE. NEXT != END
Set NODE = NODE. NEXT
        [End of Loop]
Step 5: Set NODE. NEXT = START
Step 6: Set NODE = END
Step 7: EXIT
```

The previous algorithm shows how a node is deleted from the end of the linked list. First, we check with the underflow condition. Now a node variable NODE is used to traverse the entire list until it reaches the last node of the list. In the while loop, we use another node variable PREV, which always points to the node preceding NODE. When we reach the last node and its preceding node, that is, the second to last node, we now change the next part of PREV to store the address of START. Hence, the memory occupied by the last node is freed. Finally, the second to last node now becomes the last node of the linked list. In this way, the deletion of a node from the end is done in a circular linked list.

Here is a program to implement a circular linked list.

```python
#Represents the node of list.
class Node:
    def __init__(self,data):
        self.data = data;
        self.next = None;
class CircularList:
    #Declaring head and tail pointer as null.
    def __init__(self):
        self.head = Node(None);
        self.tail = Node(None);
        self.head.next = self.tail;
        self.tail.next = self.head;
    #This function will add the new node at the end of the list.
    def add(self,data):
        newNode = Node(data);
        #Checks if the list is empty.
        if self.head.data is None:
          #If list is empty, both head and tail would point to new node.
          self.head = newNode;
            self.tail = newNode;
            newNode.next = self.head;
         else:
        #tail will point to new node.
            self.tail.next = newNode;
            #New node will become new tail.
            self.tail = newNode;
            #Since, it is circular linked list tail will point to head.
            self.tail.next = self.head;
    #Deletes node from end of the list
    def deleteEnd(self):
        #Checks whether list is empty
        if(self.head == None):
            return;
        else:
```

```
            #Checks whether contain only one element
            if(self.head != self.tail ):
                current = self.head;
                #Loop will iterate till the second last
element as current.next is pointing to tail
                while(current.next != self.tail):
                    current = current.next;
                #Second last element will be new tail
                self.tail = current;
                #Tail will point to head as it is a circular linked list
                self.tail.next = self.head;
        #If the list contains only one element
        #Then it will remove it and both head and tail will point to null
        else:
            self.head = self.tail = None;
    #Displays all the nodes in the list
    def display(self):
        current = self.head;
        if self.head is None:
            print("List is empty");
            return;
        else:
            #Prints each node by incrementing pointer.
            print(current.data),
            while(current.next != self.head):
                current = current.next;
                print(current.data),
            print("\n");
```

OUTPUT

```
>>> cll=CircularList()
>>> cll.add(7)
>>> cll.add(5)
>>> cll.add(3)
>>> cll.add(9)
>>> cll.display()
7
5
3
9


>>> cll.deleteEnd()
>>> cll.display()
7
5
3
```

Explanation: The above program has the add, deleteEnd, and display functions for the circular linked list.

- The add function add nodes at beginning of the list.
- The deleteEnd function removes node from end of the list.
- The display function prints all the nodes of the list.

4.4.5 Doubly Linked List

A *doubly linked list* is also called a two-way linked list; it is a special type of linked list that can point to the next node as well as the previous node in the sequence. In a doubly-linked list, each node is divided into three parts:

1. The first part is called the previous node, which contains the address of the previous node in the list.

2. The second part is called the information part, which contains information about the node.

3. The third part is called the next node, which contains the address of the succeeding node in the list.

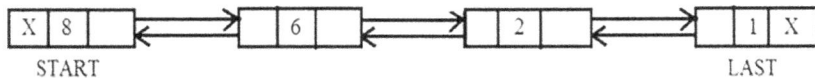

FIGURE 4.15 Doubly linked list

The structure of a doubly linked list is given as follows:

```
class ListNode :
    def __init__ ( self, data ) :
        self.data = data
        self.next = None
        self.prev = None
```

The first node of the linked list contains a NULL value in the previous node to indicate that there is no element preceding in the list; similarly, the last node also contains a NULL value in the next node field to indicate that there is no element succeeding it in the list. Doubly linked lists can be traversed in both directions.

4.4.6 Operations on a Doubly Linked List

Various operations can be performed on a circular linked list, which include

- Inserting a new node in a doubly linked list
- Deleting a node from a doubly linked list

a. Inserting a New Node in a Doubly Linked List

In this section, we learn how a new node is inserted into an already existing doubly linked list. We consider four cases for the insertion process in a doubly-linked list.

1. A new node is inserted at the beginning.

2. A new node is inserted at the end.

3. A new node is inserted after a given node.

4. A new node is inserted before a given node.

1. Inserting a new node at the beginning of a doubly-linked list

In this case of inserting a new node at the beginning of a doubly-linked list, we first check with the overflow condition, that is, whether the memory is available for a new node. If the memory is not available, then an overflow message is displayed; otherwise, the memory is allocated for the new node. Then, we initialize the node with its info part, and its address part contains the address of the first node of the list, which is the START node. Hence, the new node is added as the first node in the list, and the START node points to the first node of the list. Now to understand better, let us take an example. Consider the linked list with four nodes shown in Figure 4.16; a new node is inserted at the beginning of the linked list.

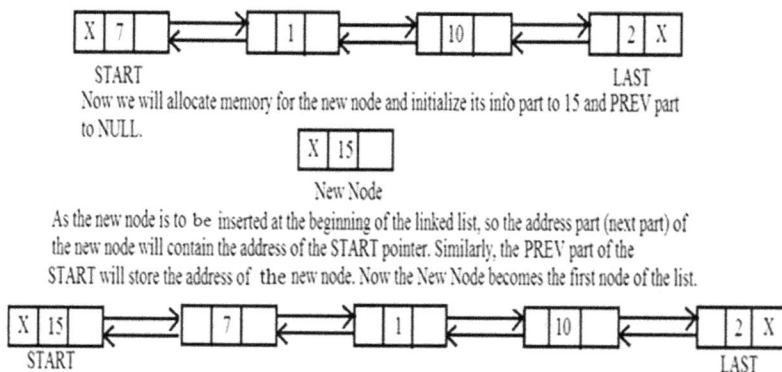

FIGURE 4.16 Inserting a new node at the beginning of a doubly linked list

From the previous example, it is clear how a new node is inserted in an already existing doubly linked list. Let us now examine its algorithm.

Algorithm for inserting a new node at the beginning of a doubly-linked list

```
Step 1: START
Step 2: IF NODE = NULL
            Print OVERFLOW
            Go to Step 9
[End of If]
Step 3: Set NEW NODE = NODE
Step 4: Set NEW NODE. INFO = VALUE
Step 5: Set NEW NODE. PREV = NULL
Step 6: Set NEW NODE. NEXT = START
Step 7: Set START. PREV = NEW NODE
Step 8: Set START = NEW NODE
Step 9: EXIT
```

2. Inserting a new node at the end of a doubly linked list

To insert the new node at the end of the linked list, we first check the overflow condition, which is to see whether the memory is available for a new node. If the memory is not available, then an overflow message is printed; otherwise, the memory is allocated for the new node. Then a NODE variable is made which initially points to START, and a NODE variable is used to traverse the list until it reaches the last node. When it reaches the last node, the NEXT part of the last node stores the address of the new node, and the NEXT part of the NEW NODE contains NULL, which denotes the end of the linked list. The PREV part of the NEW NODE stores the address of the node pointed to by NODE. Let's take a look at the algorithm for this.

Algorithm for inserting a new node at the end of a linked list

```
Step 1: START
Step 2: IF NODE = NULL
            Print OVERFLOW
[End of If]
Step 3: Set NEW NODE = NODE
Step 4: Set NODE = NODE. NEXT
Step 5: Set NEW NODE. INFO = VALUE
Step 6: Set NEW NODE. NEXT = NULL
Step 7: Set NODE = START
Step 8: Repeat while NODE. NEXT != NULL
            Set NODE = NODE. NEXT
[End of Loop]
Step 9: Set NODE. NEXT = NEW NODE
Step 10: Set NEW NODE. PREV = NODE
Step 11: EXIT
```

From the previous algorithm, we understand how to insert a new node at the end of a doubly linked list. Now, let's look at an example. Consider a linked list with four nodes as shown in Figure 4.17; a new node will be inserted at the end of the doubly linked list:

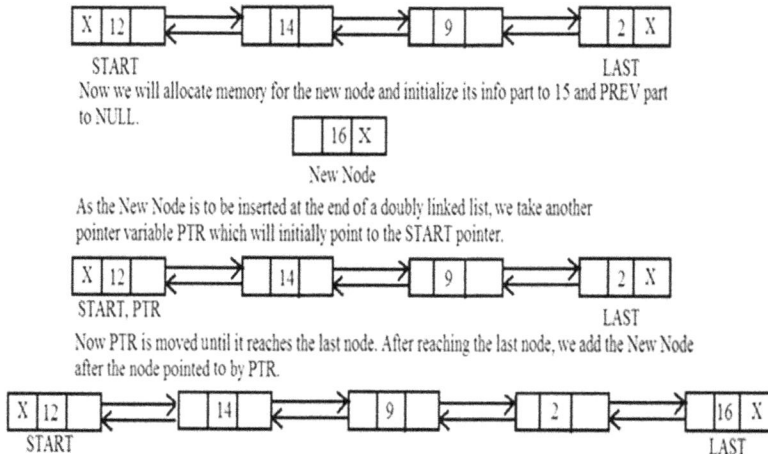

Now we will allocate memory for the new node and initialize its info part to 15 and PREV part to NULL.

As the New Node is to be inserted at the end of a doubly linked list, we take another pointer variable PTR which will initially point to the START pointer.

Now PTR is moved until it reaches the last node. After reaching the last node, we add the New Node after the node pointed to by PTR.

FIGURE 4.17 Inserting a new node at the end of a doubly linked list

3. Inserting a new node after a given node in a doubly-linked list

In this case, a new node is inserted after a given node in a doubly-linked list. As in the other cases, we again check the overflow condition in it. If the memory for the new node is available, then it is allocated; otherwise, an overflow message is displayed. Then a NODE variable is made which initially points to START, and the node variable is used to traverse the linked list until its value becomes equal to the value after which the new node is to be inserted. When it reaches that node/value, then the NEXT part of that node will store the address of the new node, and the PREV part of the NEW NODE stores the address of the preceding node. Let us examine the following algorithm.

Algorithm for inserting a new node after a given node in a linked list

```
Step 1: START
Step 2: IF NODE = NULL
            Print OVERFLOW
            Go to Step 10
        [End of If]
Step 3: Set NEW NODE = NODE
Step 4: Set NEW NODE. INFO = VALUE
Step 5: Set NODE = START
```

```
Step 6: Repeat while NODE. INFO != GIVEN_VAL
             Set NODE = NODE. NEXT
[End of Loop]
Step 7: Set NEW NODE. NEXT = NODE. NEXT
Step 8: Set NEW NODE. PREV = NODE
Step 9: Set NODE. NEXT = NEW NODE
Step 10: EXIT
```

Let's take an example. Consider a doubly linked list with four nodes as shown in Figure 4.18; a new node is inserted after a given node in the linked list.

4. Inserting a new node before a given node in a doubly-linked list

In this case, a new node is inserted before a given node in a doubly-linked list. As in the other cases, we again check the overflow condition in it. If the memory for the new node is available, then it is allocated; otherwise, an overflow message is displayed. Then a NODE variable is made which initially points to START, and the NODE variable is used to traverse the linked list until its value becomes equal to the value before which the new node is to be inserted. When it reaches that node/value, then the PREV part of that node stores the address of the NEW NODE, and the NEXT part of the NEW NODE stores the address of the succeeding node. Now to understand this better, let us take an example. Consider a linked list with four nodes as shown in Figure 4.19; a new node is inserted before a given node in the linked list.

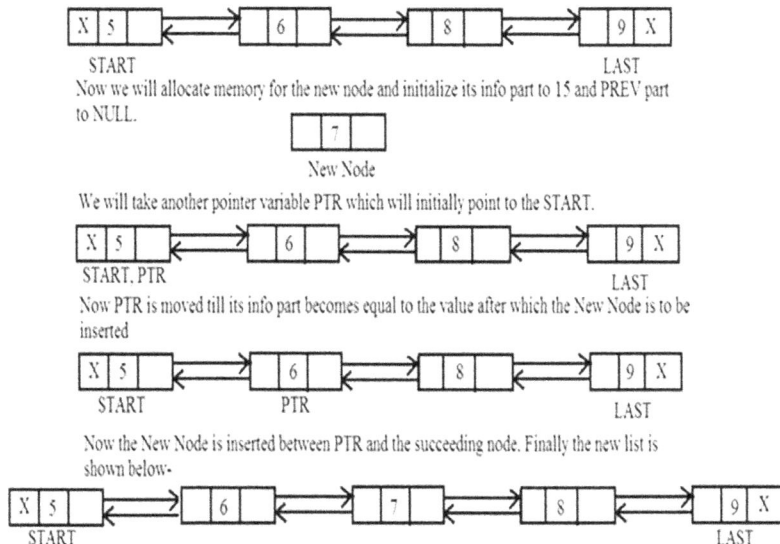

FIGURE 4.18 Inserting a new node after a given node in a doubly-linked list

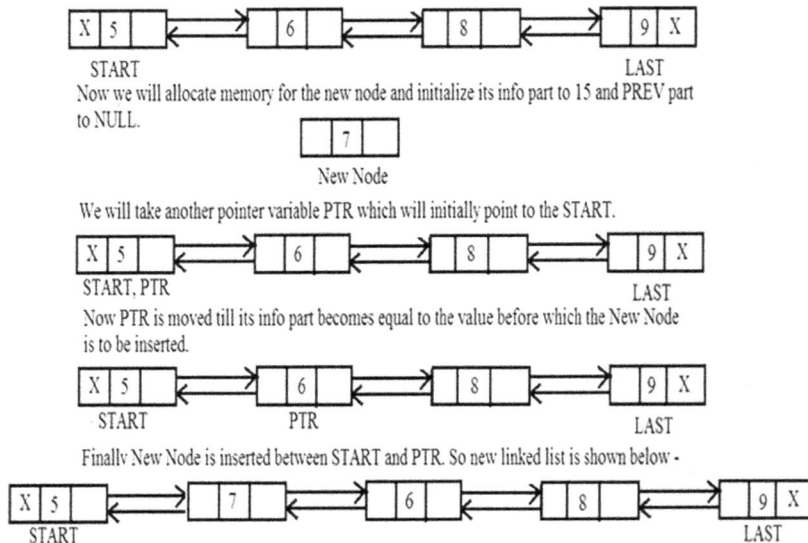

FIGURE 4.19 Inserting a new node before a given node in a doubly-linked list

From the previous example, it is clear how a new node is inserted in an already existing doubly linked list. Let us now examine its algorithm.

Algorithm for inserting a new node before a given node in a doubly-linked list

```
Step 1: START
Step 2: IF NODE = NULL
            Print OVERFLOW
            Go to Step 10
[End of If]
Step 3: Set NEW NODE = NODE
Step 4: Set NEW NODE. INFO = VALUE
Step 5: Set NODE = START
Step 6: Repeat while NODE. INFO != GIVEN_VAL
            Set NODE = NODE. NEXT
[End of Loop]
Step 7: Set NEW NODE. NEXT = NODE
Step 8: Set NEW NODE. PREV = NODE. PREV
Step 9: Set NODE. PREV = NEW NODE
Step 10: EXIT
```

b. Deleting a Node from a Doubly Linked List

In this section, we learn how a node is deleted from an already existing doubly linked list. We consider four cases for the deletion process in a doubly-linked list.

1. A node is deleted from the beginning of the linked list.

2. A node is deleted from the end of the linked list.

3. A node is deleted after a given node from the linked list.

4. A node is deleted before a given node from the linked list.

1. Deleting a node from the beginning of the doubly linked list

In the case of deleting a node from the beginning of the doubly linked list, we will first check the underflow condition, which occurs when we try to delete a node from the linked list which is empty. This situation exists when the START node is equal to NULL. Hence, if the condition is true, then the underflow message is displayed; otherwise, the node is deleted from the linked list. Consider a linked list with five nodes as shown in Figure 4.20; the node is deleted from the beginning of the linked list.

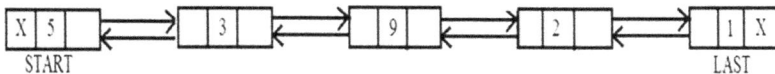

Here we will free the memory occupied by the first node of the list. Also the PREV part of the second node will contain NULL. Hence now the second node becomes the first node of the list.

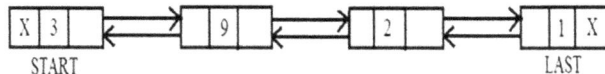

FIGURE 4.20 Deleting a node from the beginning of the doubly linked list

Let us examine an algorithm for this process.

Algorithm for deleting a node from the beginning of a doubly-linked list

```
Step 1: START
Step 2: IF START = NULL
            Print UNDERFLOW
[End Of If]
Step 3: Set NODE = START
Step 4: Set START = START. NEXT
Step 5: Set START. PREV = NULL
Step 6: FREE NODE
Step 7: EXIT
```

First, we check for the underflow condition, which is whether there are any nodes present in the linked list. If there are no nodes, then an underflow message is printed; otherwise, we move to Step 3, where we initialize NODE to START, that is, NODE now stores the address of the first node. In the next step, START is moved to the second node, as now START stores the address of the second node. Also, the PREV part of the second node now contains a value NULL. Hence, the first node is deleted and the memory that occupied NODE is freed (initially the first node of the list).

2. Deleting a node from the end of a doubly linked list

In the case of deleting a node from the end of a linked list, we first check the underflow condition. This situation exists when the START node is equal to NULL. Hence, if the condition is true, then the underflow message is printed on the screen; otherwise, the node is deleted from the linked list. Consider a linked list with five nodes as shown in Figure 4.21; the node is deleted from the end of the linked list.

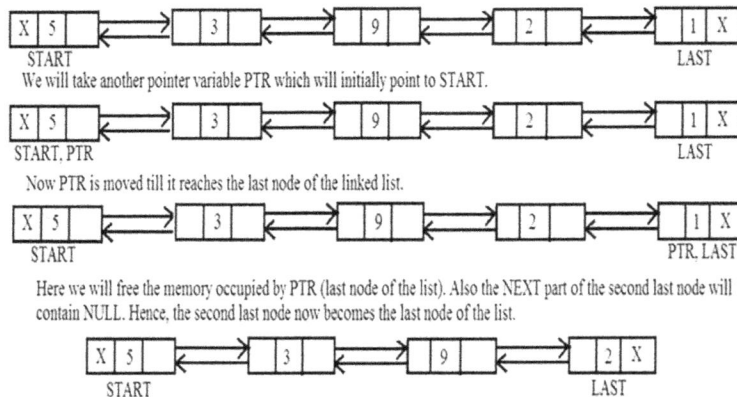

FIGURE 4.21 Deleting a node from the end of the doubly linked list

From the previous example, it is clear how a node is deleted from an already existing doubly linked list. Let us now examine its algorithm.

Algorithm for deleting a node from the end in a doubly-linked list

```
Step 1: START
Step 2: IF START = NULL
             Print UNDERFLOW
[End Of If]
Step 3: Set NODE = START
Step 4: Repeat while NODE. NEXT != NULL
             Set NODE = NODE. NEXT
[End of Loop]
Step 5: Set NODE. PREV. NEXT= NULL
Step 6: FREE NODE
Step 7: EXIT
```

We are again checking for the underflow condition. If the condition is true, then the underflow message is printed; otherwise, NODE is initialized to the START node, that is, the NODE points to the first node of the list. In the loop, the NODE is traversed until it reaches the last node of the list. After reaching the last node of the list, we can also access the second to last node by taking the address from the PREV part of the last node. Therefore, the last node is deleted, and the memory that occupied NODE is now freed.

3. Deleting a node after a given node from the doubly linked list

In the case of deleting a node after a given node from the linked list, we again check the underflow condition as we checked in both the other cases. This situation exists when the START node is equal to NULL. Hence, if the condition is true, then the underflow message is displayed; otherwise, the node is deleted from the linked list. Consider a linked list with five nodes as shown in Figure 4.22; the node is deleted after a given node from the linked list.

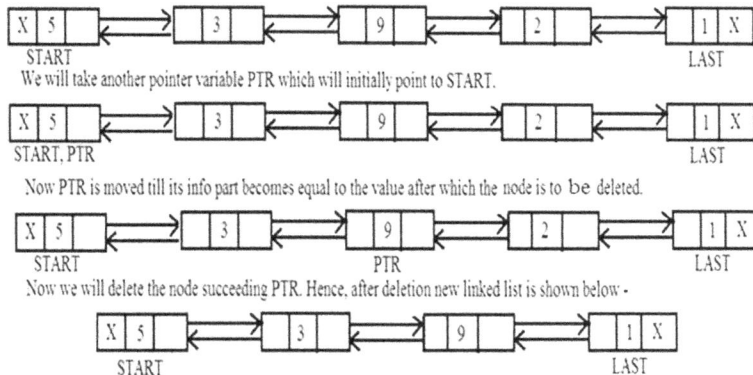

FIGURE 4.22 Deleting a node after a given node from the doubly linked list

Now let us examine this with the help of an algorithm.

Algorithm for deleting a node after a given node from the linked list

```
Step 1: START
Step 2: IF START = NULL
                Print UNDERFLOW
[End Of If]
Step 3: Set NODE = START
Step 4: Repeat while NODE. INFO != GIVEN_VAL
                Set NODE = NODE. NEXT
[End of Loop]
Step 5: Set TEMP = NODE. NEXT
Step 6: Set NODE. NEXT = TEMP. NEXT
Step 7: Set LAST. PREV = NODE
Step 8: FREE TEMP
Step 9: EXIT
```

In the algorithm, we check for the underflow condition. If the condition is true, then the underflow message is printed; otherwise, NODE is initialized to the START node, that is, the NODE points to the first node of the list. In the loop, NODE is moved until its info becomes equal to the node, after which the node is deleted. After reaching that node of the list, we can also access the succeeding node by taking the address from the NEXT part of that node. Therefore, the node is deleted and the memory is now free which had been occupied by TEMP.

4. Deleting a node before a given node from the doubly linked list

In the case of deleting a node before a given node from the linked list, we again check the underflow condition as we checked in both the other cases. This situation occurs when the START node is equal to NULL. Hence, if the condition is true, then the underflow message is printed; otherwise, the node is deleted from the linked list. Consider a linked list with five nodes as shown in Figure 4.23; the node is deleted before a given node from the linked list.

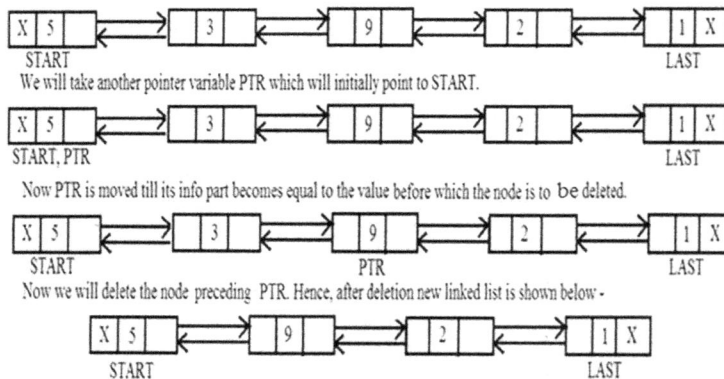

FIGURE 4.23 Deleting a node before a given node from the doubly linked list

From the previous example, it is clear how a node is deleted from an already existing doubly linked list. Let us now examine its algorithm.

Algorithm for deleting a node before a given node in a doubly-linked list

```
Step 1: START
Step 2: IF START = NULL
             Print UNDERFLOW
       [End Of If]
```

```
Step 3: Set NODE = START
Step 4: Repeat while NODE. INFO != GIVEN_VAL
            Set NODE = NODE. NEXT
[End of Loop]
Step 5: Set TEMP = NODE. PREV
Step 6: Set TEMP. PREV. NEXT = NODE
Step 7: Set NODE. PREV = TEMP. PREV
Step 8: FREE TEMP
Step 9: EXIT
```

In the previous algorithm, we check for the underflow condition. If the condition is true, then the underflow message is printed; otherwise, NODE is initialized to the START node, that is, the NODE points to the first node of the list. In the loop, NODE is moved until its info part becomes equal to the node before which the node is to be deleted. After reaching that node of the list, we can also access the preceding node by taking the address from the PREV part of that node. Therefore, the node is deleted and the memory is now free which was being occupied by TEMP.

Here is a program to implement a doubly linked list.

```python
#Represent a node of doubly linked list
class Node:
    def __init__(self,data):
        self.data = data;
        self.previous = None;
        self.next = None;
class DoublyLinkedList:
    #Represent the head and tail of the doubly linked list
    def __init__(self):
        self.head = None;
        self.tail = None;
 #addNode() will add a node to the list
 def add(self, data):
     #Create a new node
     newNode = Node(data);
     #If list is empty
     if(self.head == None):
         #Both head and tail will point to newNode
         self.head = self.tail = newNode;
         #head's previous will point to None
         self.head.previous = None;
```

```
                #tail's next will point to None, as it is the last node of the list
                self.tail.next = None;
         else:
                #newNode will be added after tail such that tail's
next will point to newNode
                self.tail.next = newNode;
                #newNode's previous will point to tail
                newNode.previous = self.tail;
                #newNode will become new tail
                self.tail = newNode;
                #As it is last node, tail's next will point to None
                self.tail.next = None;
     #deleteFromStart() will delete a node from the beginning of the list
     def delete(self):
         #Checks whether list is empty
         if(self.head == None):
              return;
         else:
                #Checks whether the list contains only one element
                if(self.head != self.tail):
                    #head will point to next node in the list
                    self.head = self.head.next;
                    #Previous node to current head will be made None
                    self.head.previous = None;
                #If the list contains only one element
                #then, it will remove the node, and now both head
and tail will point to None
                else:
                     self.head = self.tail = None;
     #display() will print out the nodes of the list
     def display(self):
         #Node current will point to head
         current = self.head;
         if(self.head == None):
              print("List is empty");
              return;
         while(current != None):
              #Prints each node by incrementing pointer.
              print(current.data),
              current = current.next;
         print();
```

OUTPUT

```
>>> dll=DoublyLinkedList()
>>> dll.add(8)
>>> dll.add(9)
>>> dll.add(2)
>>> dll.add(4)
>>> dll.display()
8
9
2
4

>>> dll.delete()
>>> dll.display()
9
2
4
```

Explanation: The above program has the add, delete and display functions for a doubly linked list.

- The add function adds nodes in beginning of the list.
- The delete function removes nodes from beginning of the list.
- The display function prints all the nodes of the list.

4.5 HEADER LINKED LISTS

Header linked lists are a special type of linked list that always contain a special node, called the header node, at the beginning. This header node usually contains vital information about the linked list, like the total number of nodes in the list, whether the list is sorted or not, and so on. There are two types of header linked lists, which include

1. **Grounded Header Linked List** – This linked list stores a unique value NULL in the address field (next part) of the last node of the list.

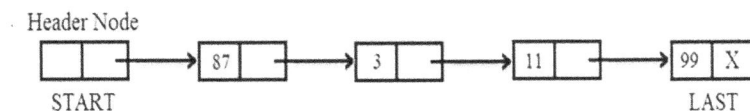

FIGURE 4.24 Grounded header linked list

2. **Circular Header Linked List** – This linked list stores the address of the header node in the address field (next part) of the last node of the list.

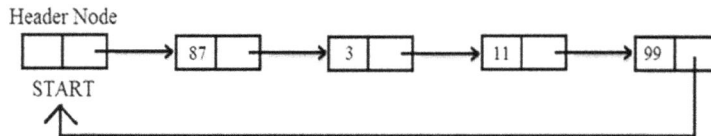

FIGURE 4.25 Circular header linked list

Frequently Asked Questions

Q3. What are the uses of a header node in a linked list?

Answer:

The header node is a node of a linked list which may or may not have the same data structure as that of a typical node. The only commonality between a typical node and a header node is that they both refer to a typical node.

4.6 APPLICATIONS OF LINKED LISTS

Linked lists have various applications, but one of the most important is that of polynomial representation; linked lists can be used to represent polynomials, and different operations can be performed on them. Now let us see how polynomials can be represented in the memory using linked lists.

4.7 POLYNOMIAL REPRESENTATION

Consider a polynomial $10x^2 + 6x + 9$. In this polynomial, every individual term consists of two parts: first, a coefficient, and second, a power. Here, the coefficients of the expression are 10, 6, and 9, and 2, 1, and 0 are the respective powers of the coefficients. Now, every individual term can be represented using a node of the linked list. The following figure shows how a polynomial expression can be represented using a linked list:

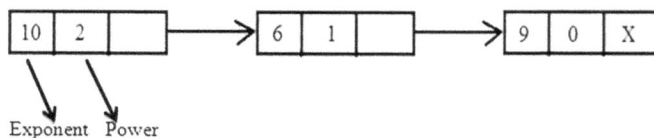

FIGURE 4.26 Linked representation of a polynomial

4.8 SUMMARY

- A linked list is a sequence of nodes in which each node contains one or more data fields and a node that points to the next node.

- The process of allocating memory during the execution of the program or the process of allocating memory to the variables at runtime is called dynamic memory allocation.

- A singly linked list is the simplest type of linked list, in which each node contains some information/data and only one node that points to the next node in the linked list.

- Traversing a linked list means accessing all the nodes of the linked list exactly once.

- Searching for a value in a linked list means to find a particular element/value in the linked list.

- A circular linked list is also a type of singly linked list in which the address part of the last node stores the address of the first node.

- A doubly linked list is also called a two-way linked list; it is a special type of linked list that can point to the next node as well as the previous node in the sequence.

- A header linked list is a special type of linked list that always contains a special node, called the header node, at the beginning. This header node usually contains vital information about the linked list like the total number of nodes in the list, whether the list is sorted or not, and so forth.

- One of the most important applications of linked lists is a polynomial representation because linked lists can be used to represent polynomials and different operations can be performed on them.

4.9 EXERCISES

4.9.1 Theory Questions

Q1. What is a linked list? How it is different from an array?

Q2. How many types of linked lists are there? Explain in detail.

Q3. What is the difference between singly and doubly linked lists?

Q4. List the various advantages of linked lists over arrays.

Q5. What is a circular linked list? What are the advantages of a circular linked list over a linked list?

Q6. Define a header linked list and explain its utility.

Q7. Give the linked representation of the following polynomial: $10x^2y - 6x + 7$.

Q8. Specify the use of a header node in a header linked list.

Q9. List the various operations that can be performed in linked lists.

4.9.2 Programming Questions

Q1. Write an algorithm/program to insert a node at the desired position in a circular linked list.

Q2. Write a Python program to insert and delete the node at the beginning in a doubly-linked list using classes.

Q3. Write an algorithm to reverse a singly linked list.

Q4. Write a Python program to delete a node from a header linked list.

Q5. Write an algorithm to concatenate two linked lists.

Q6. Write a Python program to implement a circular header linked list.

Q7. Write a Python program to count the non-zero values in a header linked list using classes.

Q8. Write a Python program that inserts a node in the linked list before a given node.

Q9. Write an algorithm to search for an element from a given linear linked list.

Q10. Write a program that inserts a node in a doubly-linked list after a given node.

4.9.3 Multiple Choice Questions

Q1. Linked lists are best suited for

a. Data structure

b. Sizes of structure and data that are constantly changing

c. Sizes of structure and data that are fixed

d. None of these

Q2. Each node in a linked list must contain at least _____ field(s).

a. Four

b. Three

c. One

d. Two

Q3. Which type of linked list stores the address of the header node in the address field of the last node?

a. Doubly linked list

b. Circular header linked list

c. Singly-linked list

d. Header linked list

Q4. The situation in a linked list when START = NULL is

a. Overflow

b. Underflow

c. Both

d. None of these

Q5. Linked lists can be implemented in what type of data structures?

a. Queues

b. Trees

c. Stacks

d. All of these

Q6. Which type of linked list contains a node to the next as well as the previous nodes?

a. Doubly linked list

b. Singly-linked list

c. Circular linked list

d. Header linked list

Q7. The first node in the linked list is called _____.

    a. End

    b. Middle

    c. Start

    d. Begin

Q8. A linked list cannot grow and shrink during compile time.

    a. False

    b. It might grow

    c. True

    d. None of the above

Q9. What does NULL represent in the linked list?

    a. Start of list

    b. End of list

    c. None of the above

## CHAPTER 5

QUEUES

5.1 INTRODUCTION

A queue is an important data structure that is widely used in many computer applications. A queue can be visualized with many examples from everyday life. A very simple illustration of a queue is a line of people standing outside a movie theater. The first person standing in the line will enter the movie theatre first. We observe that whenever we talk about a queue, we see that that the element in the first position is served first. Thus, a queue can be described as a FIFO (First-in, First-out) data structure; that is, the element that is inserted first will be the first one to be taken out.

5.2 DEFINITION OF A QUEUE

A queue is a linear collection of data elements in which the element inserted first is the element taken out first (i.e., a queue is a FIFO data structure). A queue is an abstract data structure, somewhat similar to stacks. Unlike stacks, a queue is open on both ends. A *queue* is a linear data structure in which the first element is inserted on one end, called the REAR end (also called the tail end), and the deletion of the element takes place from the other end, called the FRONT end (also called the head). One end is always used to insert data and the other end is used to remove data.

Queues can be implemented by using arrays or linked lists. We discuss the implementation of queues using arrays and linked lists in this section.

Practical Application:

- A real-life example of a queue is people moving on an escalator. The people who got on the escalator first will be the first ones to step off of it.
- Another illustration of a queue is a line of people standing at the bus stop waiting for the bus. Therefore, the first person standing in the line will get into the bus first.

5.3 IMPLEMENTATION OF A QUEUE

Queues can be implemented using two data structures:

1. arrays/lists

2. linked lists

5.3.1 Implementation of Queues Using Arrays

Queues can be easily implemented using arrays. Initially, the front end (head) and the rear end (tail) of the queue point at the first position or location of the array. As we insert new elements into the queue, the rear keeps on incrementing, always pointing to the position where the next element will be inserted, while the front remains in the first position.

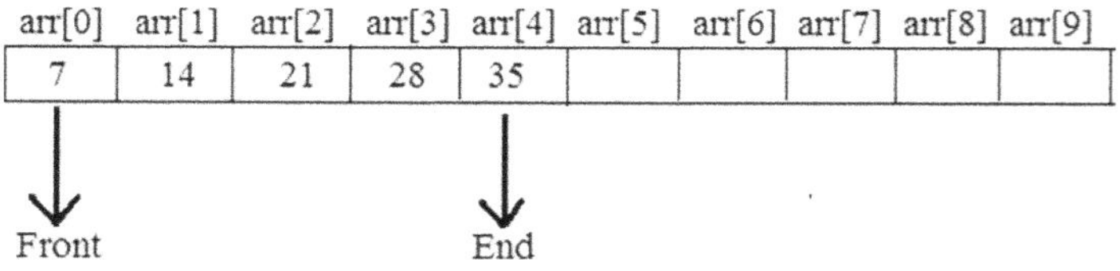

FIGURE 5.1 Array representation of a queue

5.3.2 Implementation of Queues Using Linked Lists

We have already studied how a queue is implemented using an array. Now let us discuss the same using linked lists. We already know that in linked lists, dynamic memory allocation takes place; that is, the memory is allocated at runtime. But in the case of arrays, memory is allocated at the start of the program. (We discussed in the chapter about linked lists.) If we are aware of the maximum size of the queue in advance, then the implementation of a queue using arrays is efficient. But if the size is not known in advance, then we use the concept of a linked list, in which dynamic memory allocation takes place. A linked list has two parts: the first part contains the information of the node, and the second part stores the address of the next element in the linked list. Similarly, we can also implement a linked queue. The START node in the linked list becomes the FRONT node in a linked queue, and the end of the queue is denoted by REAR. All insertion operations are done at the rear end only. Similarly, all deletion operations are done at the front end only.

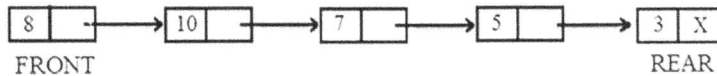

FIGURE 5.2 A linked queue

5.3.2.1 Insertion in Linked Queues

Insertion is the process of adding new elements in the already existing queue. The new elements in the queue are always inserted from the rear end. Initially, we check whether FRONT = NULL. If the condition is true, then the queue is empty; otherwise, the new memory is allocated for the new node. We can examine this further with the help of an algorithm.

Algorithm for Inserting a New Element in a Linked Queue

```
Step 1: START
Step 2: Set NEW NODE . INFO = VAL
IF FRONT = NULL
          Set FRONT = REAR = NEW NODE
          Set FRONT . NEXT = REAR . NEXT = NEW NODE
ELSE
          Set REAR . NEXT = NEW NODE
          Set NEW NODE . NEXT = NULL
          Set REAR  = NEW NODE
 [End of If]
Step 3: EXIT
```

First, we allocate the memory for the new node. Then we initialize it with the information to be stored in it. Next, we check if the new node is the first node of the queue or not. If the new node is the first node of the queue, then we store NULL in the address part of the new node. In this case, the new node is tagged as FRONT as well as REAR. However, if the new node is not the first node of the queue, it is inserted at the REAR end of the queue.

For Example – Consider a linked queue with five elements; a new element is to be inserted in the queue.

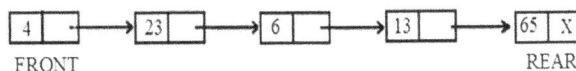

FIGURE 5.3 Linked queue before insertion

After inserting the new element in the queue, the updated queue becomes as shown in Figure 5.4.

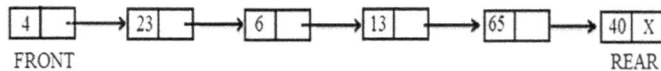

FRONT   REAR

FIGURE 5.4 Linked queue after insertion.

5.3.2.2 Deletion in Linked Queues

Deletion is the process of removing elements from the already existing queue. The elements from the queue will always be deleted from the front end. Initially, we check with the underflow condition, that is, whether FRONT = NULL. If the condition is true, then the queue is empty, which means we cannot delete any elements from it. Therefore, in that case, an underflow error message is displayed on the screen. We can examine this further with the help of an algorithm.

Algorithm for Deleting an Element from a Queue

```
Step 1: START
Step 2: IF FRONT = NULL
            Print UNDERFLOW ERROR
[End of If]
Step 3: Set TEMP = FRONT
Step 4: Set FRONT = FRONT . NEXT
Step 5: FREE TEMP
Step 6: EXIT
```

We first check with the underflow condition, that is, whether the queue is empty or not. If the condition is true, then an underflow error message is displayed; otherwise, we use a node variable TEMP that points to FRONT. In the next step, FRONT is now pointing to the second node in the queue. Finally, the first node is deleted from the queue.

For Example – Consider a linked queue with five elements; an element is to be deleted from the queue.

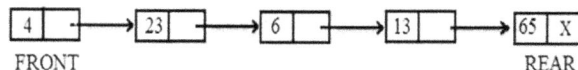

FRONT   REAR

After deleting an element from the queue, the updated queue becomes as shown in Figure 5.5.

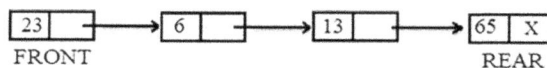

FRONT   REAR

FIGURE 5.5 Linked queue after deletion

Here is a program for implementing a linked queue performing insertion and deletion operations.

```python
# Python3 program to demonstrate a linked list
# based implementation of queue

# A linked list (LL) node
# to store a queue entry
class Node:

    def __init__(self, data):
        self.data = data
        self.next = None

# A class to represent a queue

# The queue, front stores the front node
# of LL and rear stores the last node of LL
class Queue:

    def __init__(self):
        self.front = self.rear = None

    def isEmpty(self):
        return self.front == None

    # Method to add an item to the queue
    def insert(self, item):
        temp = Node(item)

        if self.rear == None:
            self.front = self.rear = temp
            return
        self.rear.next = temp
        self.rear = temp

    # Method to remove an item from queue
    def Delete(self):

        if self.isEmpty():
            return
        temp = self.front
        self.front = temp.next

        if(self.front == None):
            self.rear = None
    #method to print queue
    def display(self):
        if self.isEmpty():
            return
        temp=self.front
        while(temp):
            print(temp.data)
            temp=temp.next
```

The output of the program is

```
>>> q=Queue()
>>> q.insert(5)
>>> q.display()
5
>>> q.insert(6)
>>> q.insert(7)
>>> q.insert(9)
>>> q.display()
5
6
7
9
>>> q.Delete()
>>> q.display()
6
7
9
>>> q.Delete()
>>> q.display()
7
9
```

Explanation: The above program has the insert, delete, and display functions for a linked queue.

- The insert function adds an element to the queue.
- The delete function removes elements from the queue.
- The display function prints every node of the queue.

Frequently Asked Questions

Q. Define queues; in what ways can a queue be implemented?

Answer:

A queue is a linear data structure in which the first element is inserted from one end, called the REAR end (also called the tail end), and the deletion of the element takes place from the other end called the FRONT end (also called the head). Each type of queue can be implemented in two ways:

1. *Array/List Representation*
2. *Linked List Representation*

5.4 OPERATIONS ON QUEUES

The two basic operations that can be performed on queues are as follows.

5.4.1 Insertion

Insertion is the process of adding new elements in the queue. However, before inserting any new element in the queue, we must always check for the overflow condition, which occurs when we try to insert an element in a queue that is already full. An overflow condition can be checked as follows: If REAR = MAX − 1, where MAX is the size of the queue. Hence, if the overflow condition is true, then an overflow message is displayed on the screen; otherwise, the element is inserted into the queue. Insertion is always done at the rear end. Insertion is also known as en-queue.

For Example – Let us take a queue that has five elements in it. Suppose we want to insert another element, 50, in it; then REAR will be incremented by 1. Thus, a new element is inserted at the position pointed to by REAR. Now, let us see how insertion is done in the queue in Figure 5.6.

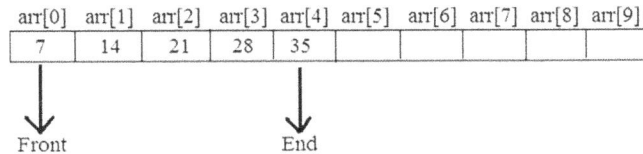

arr[0]	arr[1]	arr[2]	arr[3]	arr[4]	arr[5]	arr[6]	arr[7]	arr[8]	arr[9]
7	14	21	28	35					

Front   End

After inserting 50 in it, the new queue is

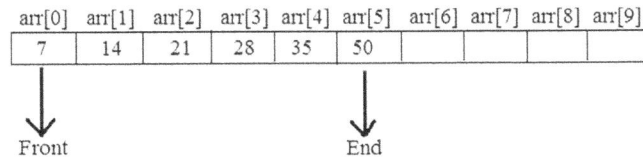

arr[0]	arr[1]	arr[2]	arr[3]	arr[4]	arr[5]	arr[6]	arr[7]	arr[8]	arr[9]
7	14	21	28	35	50				

Front   End

FIGURE 5.6 Queue after inserting a new element

Algorithm for Inserting a New Element in a Queue

```
Step 1: START
Step 2: IF REAR = MAX - 1
            Print OVERFLOW ERROR
[End of If]
Step 3: IF FRONT = -1 && REAR = -1
            Set FRONT = 0
            Set REAR = 0
            ELSE
            REAR = REAR + 1
[End of If]
Step 4: Set QUE[REAR] = ITEM
Step 5: EXIT
```

In the previous algorithm, we first check for the overflow condition. In Step 2, we check to see whether the queue is empty. If the queue is empty, then both FRONT and REAR are set to zero; otherwise, REAR is incremented to the next position in the queue. Finally, the new element is stored in the queue at the position pointed to by REAR.

5.4.2 Deletion

Deletion is the process of removing elements from the queue. However, before deleting any element from the queue, we must always check for the underflow condition, which occurs when we try to delete an element from a queue that is empty. An underflow condition can be checked as follows: If FRONT > REARorFRONT = −1. Hence, if the underflow condition is true, then an underflow message is displayed on the screen; otherwise, the element is deleted from the queue. Deletion is always done at the front end. Deletion is also known as de-queue.

For Example – Let us take a queue with five elements in it. Suppose we want to delete an element, 7, from a queue; then FRONT will be incremented by 1. Thus, the new element is deleted from the position pointed to by FRONT. Now, let us see how the deletion is done in the queue in Figure 5.7.

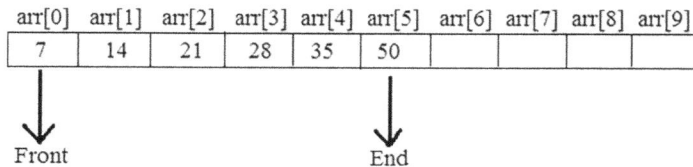

arr[0]	arr[1]	arr[2]	arr[3]	arr[4]	arr[5]	arr[6]	arr[7]	arr[8]	arr[9]
7	14	21	28	35	50				

Front ↓   End ↓

After deleting 7 from it, the new queue will be

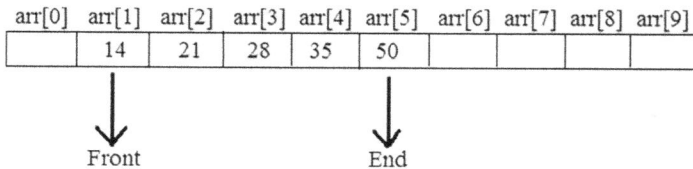

arr[0]	arr[1]	arr[2]	arr[3]	arr[4]	arr[5]	arr[6]	arr[7]	arr[8]	arr[9]
	14	21	28	35	50				

Front ↓   End ↓

FIGURE 5.7 Queue after deleting an element

Algorithm for Deleting an Element from a Queue

```
Step 1: START
Step 2: IF FRONT > REAR or FRONT = -1
            Print UNDERFLOW ERROR
[End of If]
```

```
Step 3: Set ITEM = QUE[FRONT]
Step 4:Set FRONT = FRONT + 1
Step 5: EXIT
```

First, we check for the underflow condition, that is, whether the queue is empty or not. If the queue is empty, then no deletion takes place; otherwise, the FRONT is incremented to the next position in the queue. Finally, the element is deleted from the queue.

Here is a menu-driven program for a linear queue performing insertion and deletion operations.

```python
# Python program to
# demonstrate queue implementation
# using list

# Initializing a queue
queue = []

#function to display queue
def display():
    print(queue)
#function to delete element from queue
def delete():
    temp=queue[0]
    queue.pop(0)
    print(temp,"is deleted")
#function to insert element in queue
def insert():
    data=input("enter data to be insert-")
    queue.append(data)
    print("success")
#menu for queue operations
while(1):
    print("    menu    ")
    print("1-insert")
    print("2-delete")
    print("3-display")
    print("4-exit")
    choice=input("enter choice-")
    if choice=="1":
        insert()
    elif choice=="2":
        delete()
    elif choice=="3":
        display()
    elif choice=="4":
        exit(0)
```

The output of the program is

```
    menu
1-insert
2-delete
3-display
4-exit
enter choice-1
enter data to be insert-3
success
    menu
1-insert
2-delete
3-display
4-exit
enter choice-1
enter data to be insert-5
success
    menu
1-insert
2-delete
3-display
4-exit
enter choice-1
enter data to be insert-7
success
```

```
    menu
1-insert
2-delete
3-display
4-exit
enter choice-3
['3', '5', '7']
    menu
1-insert
2-delete
3-display
4-exit
enter choice-2
3 is deleted
    menu
1-insert
2-delete
3-display
4-exit
enter choice-3
['5', '7']
```

Explanation: The above menu-driven program has the insert, delete, and display functions for a linear queue.

- The insert function adds an element to the queue.
- The delete function removes an element from the queue.
- The display function prints every node of the queue.

5.5 TYPES OF QUEUES

This section discusses various types of queues which include

1. Circular Queue

2. Priority Queue

3. De-Queue (Double-ended Queue)

5.5.1 Circular Queue

A *circular queue* is a special type of queue implemented in a circular fashion rather than in a straight line. A circular queue is a linear data structure in which the operations are performed based on the FIFO principle and the last position is connected to the first position to make a circle. It is also called a "ring buffer."

5.5.1.1 Limitation of Linear Queues

In linear queues, we studied how insertion and deletion take place. We discussed that inserting a new element in the queue is only done at the rear end. Similarly, deleting an element from the queue is only done at the front end. Now let us consider a queue of 10 elements.

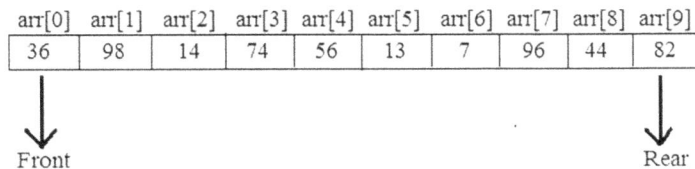

arr[0]	arr[1]	arr[2]	arr[3]	arr[4]	arr[5]	arr[6]	arr[7]	arr[8]	arr[9]
36	98	14	74	56	13	7	96	44	82

Front ↓   Rear ↓

The queue is now full, so we cannot insert any more elements in it. If we delete three elements from the queue, the queue will be as follows:

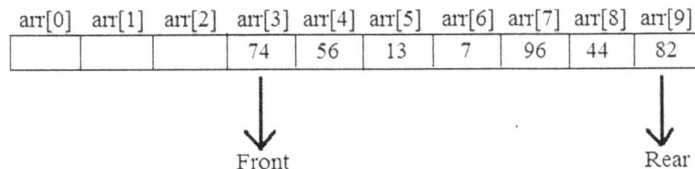

arr[0]	arr[1]	arr[2]	arr[3]	arr[4]	arr[5]	arr[6]	arr[7]	arr[8]	arr[9]
			74	56	13	7	96	44	82

Front ↓   Rear ↓

Thus, we can see that even after the deletion of three elements from the queue, the queue is still full, as REAR = MAX – 1. We still cannot insert any new elements in it as there is no space to store new elements. This is a major drawback of the linear queue.

To overcome this problem, we shift all the elements to the left so that the new elements can be inserted from the rear end, but shifting all the elements of the queue can be a very time-consuming procedure, as queues are typically very large. Another solution to this problem is a circular queue. First, let us see how a circular queue looks (Figure 5.8).

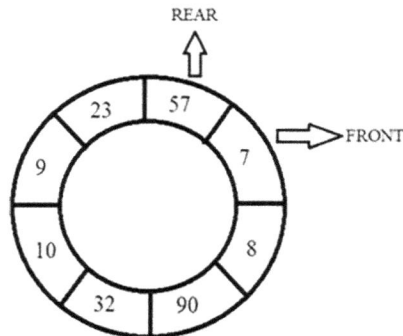

FIGURE 5.8 A circular queue

In a circular queue, the elements are stored in a circular form such that the first element is next to the last element in the queue. A circular queue will be full when FRONT = 0 and REAR = MAX – 1 or FRONT = REAR + 1. In that case, an overflow error message will be displayed on the screen. Similarly, a circular queue is empty when both FRONT and REAR are equal to zero. In that case, an underflow error message is displayed on the screen. Now, let us study both insertion and deletion operations in a circular queue.

Practical Application:

A circular queue is used in operating systems for scheduling different processes.

Frequently Asked Questions

Q. What is a circular queue? List the advantages of a circular queue over a simple queue.

Answer:

A circular queue is a particular kind of queue where new items are added to the rear end of the queue and items are read off from the front end of the queue, so there is a constant stream of data flowing in and out of the queue. A circular queue is also known as a "circular buffer." It is a structure that allows data to be passed from one process to another, making the most efficient use of memory. The only difference between a linear queue and circular queue is that in a linear queue, when the rear points to the last position in the array, we cannot insert data even if we have deleted some elements. But in a circular queue, we can insert elements as long as there is free space available. The main advantage of a circular queue as compared to a linear queue is that it avoids wasting space.

5.5.1.2 Inserting an Element in a Circular Queue

While inserting a new element in the already existing queue, we first check for the overflow condition, which occurs when we are trying to insert an element in the queue that is already full. The position of the new element to be inserted can be calculated by using the following formula:

REAR = (REAR + 1) % MAX, where MAX is equal to the size of the queue.

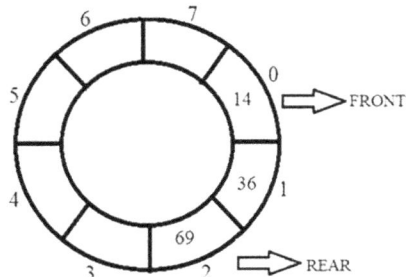

FIGURE 5.9 Initial circular queue without insertion

For Example – Let us consider a circular queue with three elements in it. Suppose we want to insert an element, 56. Let us see how insertion is done in the circular queue.

Step 1: Initially the queue contains three elements. FRONT denotes the beginning of the circular queue, and REAR denotes the end of the circular queue.

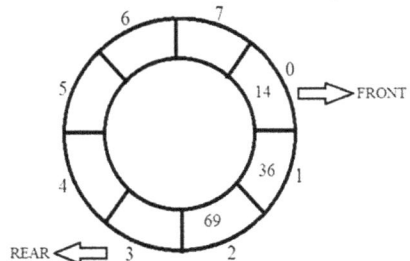

FIGURE 5.10 REAR is incremented by 1 so that it points to the next location

Step 2: Now, the new element is to be inserted in the queue. Hence, REAR = REAR + 1; that is, REAR will be incremented by 1 so that it points to the next location in the queue.

Step 3: Finally, in this step, the new element is inserted at the location pointed to by REAR. Hence, after insertion the queue is as shown in Figure 5.11.

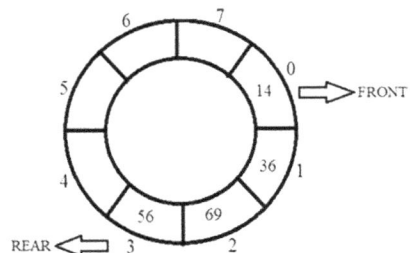

FIGURE 5.11 Final queue after inserting a new element

Algorithm for Inserting an Element in a Circular Queue

Here, QUEUE is an array with N elements. FRONT and REAR point to the front and rear elements of the queue. ITEM is the value to be inserted.

```
Step 1: START
Step 2:IF (FRONT = 0 && REAR = MAX - 1) OR (FRONT = REAR + 1)
            Print OVERFLOW ERROR
Step 3: ELSE
            IF (FRONT = -1)
            Set FRONT = 0
            Set REAR = 0
Step 4:ELSE
            IF (REAR = MAX - 1)
            Set REAR = 0
            ELSE
            REAR = REAR + 1
[End of If]
[End of If]
Step 5:Set CQUEUE[REAR] = ITEM
Step 6: EXIT
```

First, we check with the overflow condition. Second, we check if the queue is empty. If the queue is empty, then the FRONT and REAR are set to zero. In Step 4, if REAR has reached its maximum capacity, then we set REAR = 0; otherwise, REAR is incremented by 1 so that it points to the next position where the new element is to be inserted. Finally, the new element is inserted in the queue.

5.5.1.3 Deleting an Element from a Circular Queue

While deleting an element from the already existing queue, we first check for the underflow condition, which occurs when we are trying to delete an element from a queue that is empty. After deleting an element from the circular queue, the position of the FRONT end can be calculated by the following formula:

FRONT = (FRONT + 1) % MAX, where MAX is equal to the size of the queue.

For Example – Let us consider a circular queue with seven elements in it. Suppose we want to delete an element, 45, from it. Let us see how the deletion is done in the circular queue.

Step 1: Initially, the queue contains seven elements. FRONT denotes the beginning of the circular queue, and REAR denotes the end of the circular queue.

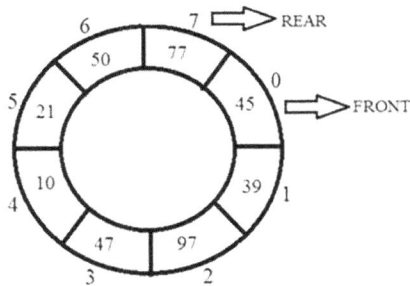

FIGURE 5.12 Initial circular queue without deletion

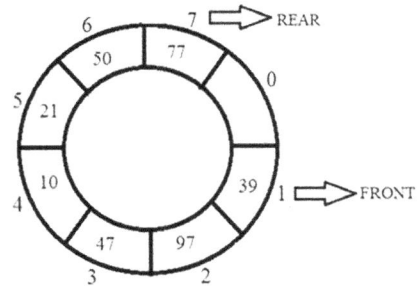

FIGURE 5.13 Final queue after deleting an element

Step 2: Now, the element is to be deleted from the queue. Hence, FRONT = FRONT + 1, that is, FRONT will be incremented by 1 so that it points to the next location in the queue. Also, the value is deleted from the queue. Thus, the queue after deletion is shown in Figure 5.13.

Algorithm for Deleting an Element from a Circular Queue

Here, CQUEUE is an array with N elements. FRONT and REAR point to the front and rear elements of the queue. ITEM is the value to be deleted.

```
Step 1: START
Step 2: IF (FRONT = -1)
            Print UNDERFLOW ERROR
Step 3: ELSE
            Set ITEM = CQUEUE[FRONT]
Step 4:IF (FRONT = REAR)
            Set FRONT = -1
            Set REAR = -1
Step 5:ELSE IF (FRONT = MAX - 1)
            Set FRONT = 0
            ELSE
            FRONT = FRONT + 1
[End of If]
[End of If]
Step 6: EXIT
```

We first check with the underflow condition. Second, we store the element to be deleted in ITEM. Third, we check to see if the queue is empty or not after deletion. If FRONT has reached its maximum capacity, then we set FRONT = 0; otherwise, the FRONT is incremented by 1 so that it points to the next position. Finally, the element is deleted from the queue.

Here is a program for a linear circular queue performing insertion and deletion operations.

```python
class CircularQueue():

    # constructor
    def __init__(self, size): # initializing the class
        self.size = size

        # initializing queue with none
        self.queue = [None for i in range(size)]
        self.front = self.rear = -1

    def insert(self, data):

        # condition if queue is full
        if ((self.rear + 1) % self.size == self.front):
            print(" Queue is Full\n")

        # condition for empty queue
        elif (self.front == -1):
            self.front = 0
            self.rear = 0
            self.queue[self.rear] = data
        else:

            # next position of rear
            self.rear = (self.rear + 1) % self.size
            self.queue[self.rear] = data

    def delete(self):
        if (self.front == -1): # condition for empty queue
            print ("Queue is Empty\n")

        # condition for only one element
        elif (self.front == self.rear):
            temp=self.queue[self.front]
            self.front = -1
            self.rear = -1
            return temp
        else:
            temp = self.queue[self.front]
            self.front = (self.front + 1) % self.size
            return temp

    def display(self):

        # condition for empty queue
        if(self.front == -1):
            print ("Queue is Empty")

        elif (self.rear >= self.front):
            print("Elements in the circular queue are:",
                                                end = " ")
```

```
            for i in range(self.front, self.rear + 1):
                print(self.queue[i], end = " ")
            print ()
        else:
            print ("Elements in Circular Queue are:",
                                           end = " ")
            for i in range(self.front, self.size):
                print(self.queue[i], end = " ")
            for i in range(0, self.rear + 1):
                print(self.queue[i], end = " ")
            print ()

        if ((self.rear + 1) % self.size == self.front):
            print("Queue is Full")
```

The output of the program is

```
>>> cq=CircularQueue(10)
>>> cq.insert(3)
>>> cq.insert(6)
>>> cq.insert(9)
>>> cq.insert(5)
>>> cq.display()
Elements in the circular queue are: 3 6 9 5
>>> cq.delete()
3
>>> cq.display()
Elements in the circular queue are: 6 9 5
>>>
```

Explanation: The above program has the insert, delete, and display functions for a linear circular queue.

- The insert function adds element to the circular queue.
- The delete function removes element from the circular queue.
- The display function prints every node of the circular queue.

5.5.2 Priority Queue

A *priority queue* is another variant of a queue in which elements are processed based on the assigned priority. Each element in a priority queue is assigned a special value called the *priority* of the element. The elements in the priority queue are processed based on the following rules:

1. An element with a higher priority is processed first, and then the element with lower priority is processed.

2. If the two elements have the same priority, then the elements are processed on the First Come, First Served (FCFS) basis. The priority of the element is selected by its value, called the implicit priority, and the priority number given with each element is called the explicit priority.

A priority queue is like a modified queue or stack data structure, but where additionally each element has a priority associated with it. In a priority queue, the insertion and deletion operations are also done according to the assigned priority. If we want to delete an element from the priority queue, then the element with the highest priority is processed first and is deleted. The case is the same with insertion. The priority given to the elements in the queue is based on several factors. Priority queues are commonly used in operating systems for executing higher priority processes first. The priority assigned to these processes may be based on the time taken by the CPU to execute these processes completely.

Practical Application:

In an operating system, if there are four processes to be executed where the first process needs 3 ns to complete, the second process needs 5 ns to complete, the third process needs 9 ns to complete, and the fourth needs 8 ns to complete, then the first process will be given the highest priority and will be the first to be executed among all the processes.

Now the priority queues are further divided into two types which are

1. **Ascending Priority Queue** – In this type of priority queue, elements can be inserted in any order, but at the time of the deletion of elements from the queue, the smallest element is searched and deleted first.

2. **Descending Priority Queue** – In this type of priority queue, elements can be inserted in any order. But at the time of the deletion of elements from the queue, the largest element is searched and deleted first (for example – Operating systems, Routing).

Frequently Asked Questions

Q. Define Priority Queue.

Answer:

A priority queue is a collection of elements such that each element has been assigned a priority and such that the order in which elements are deleted and processed comes from the following rules:

a. *An element of higher priority is processed before any element of lower priority.*

b. *Two elements with the same priority are processed according to the order in which they were added to the queue.*

The array elements in a priority queue can have the following structure:

```
class ListNode :
      def __init__ ( self, data ) :
            self.priority=priority
      self.data = data
      self.next = None
```

5.5.2.1 Implementation of a Priority Queue

A priority queue can be implemented in two ways:

1. Array Representation of a Priority Queue

2. Linked Representation of a Priority Queue

Let us now discuss both these implementations in detail.

1. Implementation of a priority queue using arrays

While implementing a priority queue using arrays, the following points must be considered:

- Maintain a separate queue for each level of priority or priority number.
- Each queue will appear in its circular array and must have its pairs of nodes, that is, FRONT AND REAR.
- If each queue is allocated the same amount of memory, then a 2D array can be used instead of a linear array.

For Example – FRONT [K] and REAR [K] are the nodes containing the front and rear values of row "K" of the queue, where K is the priority number. If we want to insert an element with priority K, then we will add the element at the REAR end of row K; K is the row as well as the priority number of that element. If we add F with priority number 4, then the queue will be given as shown in the following table.

FRONT	REAR
2	2
1	3
0	0
5	1
4	4

FIGURE 5.14 Priority queue after inserting a new element

2. Implementation of a priority queue using linked lists

A priority queue can be implemented using a linked list. When implementing the priority queue using a linked list, every node has three parts:

a. Information part

b. Priority number of the element

c. Address of the next element

An element with higher priority precedes the element having a lower priority. Also, the priority number and priority are opposite to each other; that is, an element having a lower priority number means it has higher priority. For example, if there are two elements, X and Y, with priority numbers 2 and 7, respectively, then X will be processed first because it has a higher priority.

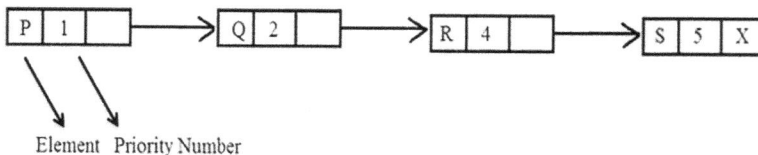

FIGURE 5.15 A linked priority queue

5.5.2.2 Insertion in a Linked Priority Queue

While inserting a new element in a linked priority queue, we traverse the entire queue until we find a node that has a lower priority than the new element. Thus, the new element is inserted before the element with the lower priority. If there is an element in the queue that has the same priority as that of the new element, then the new element is inserted after that element.

For Example – Consider a priority queue with four elements given as follows:

FIGURE 5.16 Linked priority queue before insertion

Now, a new element with information A and priority number 3 is to be inserted; hence, the element is inserted before R that has priority number 4, which is lower than that of the new element. The priority queue after inserting a new element is shown in Figure 5.17.

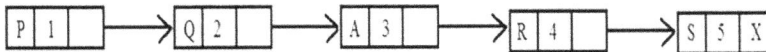

FIGURE 5.17 Linked priority queue after inserting a new element

5.5.2.3 Deletion in a Linked Priority Queue

Deleting an element from a linked priority queue is a very simple process. The first node from the priority queue is deleted and the information of that node is processed first.

For Example – Consider a priority queue with five elements given as follows:

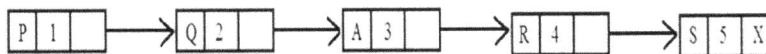

FIGURE 5.18 Linked priority queue before deletion

Now, the first node from the queue is deleted. So, the priority queue after deletion is shown as follows:

FIGURE 5.19 Linked priority queue after deleting the first node

Here is a program for a priority queue performing insertion and deletion operations.

```python
# class for Node with data and priority
class Node:

  def __init__(self, info, priority):
    self.info = info
    self.priority = priority
# class for Priority queue
class PriorityQueue:

  def __init__(self):
    self.queue = list()
    # if you want you can set a maximum size for the queue

  def insert(self, node):
    # if queue is empty
    if self.size() == 0:
      # add the new node
      self.queue.append(node)
    else:
      # traverse the queue to find the right place for new node
      for x in range(0, self.size()):
        # if the priority of new node is greater
        if node.priority >= self.queue[x].priority:
          # if we have traversed the complete queue
          if x == (self.size()-1):
            # add new node at the end
            self.queue.insert(x+1, node)
          else:
            continue
        else:
          self.queue.insert(x, node)
          return True

  def delete(self):
    # remove the first node from the queue
    return self.queue.pop(0)

  def show(self):
    for x in self.queue:
      print (str(x.info)+" - "+str(x.priority))

  def size(self):
    return len(self.queue)
```

The output of the program is

```
>>> pq=PriorityQueue()
>>> node1=Node(12,5)
>>> node2=Node(45,3)
>>> node3=Node(23,1)
>>> node4=Node(4,4)
>>> node5=Node(33,2)
>>> pq.insert(node1)
```

```
>>> pq.insert(node2)
True
>>> pq.insert(node3)
True
>>> pq.insert(node4)
True
>>> pq.insert(node5)
True
>>> pq.show()
23 - 1
33 - 2
45 - 3
4 - 4
12 - 5
>>> pq.delete()
<__main__.Node object at 0x03CD73B8>
>>> pq.show()
33 - 2
45 - 3
4 - 4
12 - 5
```

Explanation: The above program has the insert, delete and show functions for a priority queue.

- The insert function adds an element with assigned priority to the priority queue.
- The delete function removes the highest priority element from the priority queue.
- The show function prints every element with their assigned priority of the priority queue.

5.5.3 De-queues (Double-Ended Queues)

A Double-Ended queue (de-queue, pronounced "deck") is a special type of data structure in which the insertion and deletion of elements are done at either end, that is, either at the front end or at the rear end of the queue. It is often called a head-tail linked list because the elements are added or removed from either the head (front) end or tail (end). De-queues are implemented using circular arrays in the computer's memory. LEFT and RIGHT are maintained in the de-queue, which point to either end of the queue.

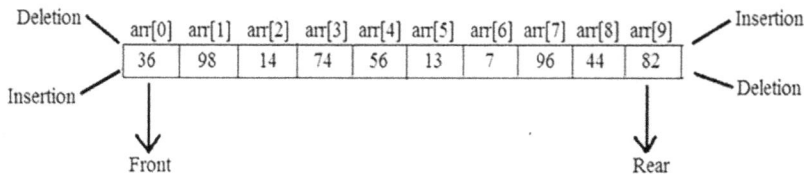

FIGURE 5.20 A double-ended queue

Practical Application:

A real-life example of a de-queue is that of a train station, where the entry and exit of passengers can take place from both sides.

There are two types of double-ended queues, which include

1. **Input Restricted De-Queue** – In this, the deletion operation can be performed at both ends (i.e., both the front and rear end) while the insertion operation can be performed only at one end (i.e., rear-end).

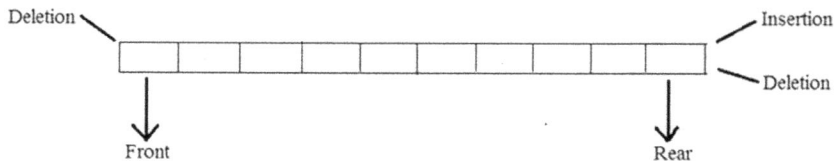

FIGURE 5.21 An input restricted a double-ended queue

2. **Output Restricted De-Queue** – In this, the insertion operation can be performed at both ends, while the deletion operation can be performed only at one end (i.e., the front end).

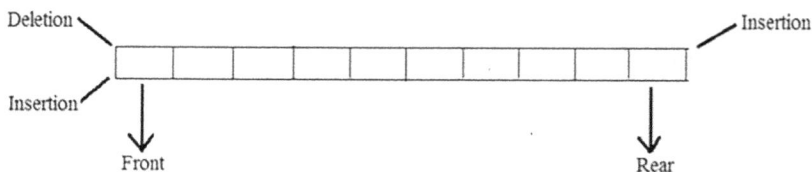

FIGURE 5.22 An output restricted a double-ended queue

Here is a menu-driven program for a double-ended queue performing the insertion and deletion operations.

```python
# Python program to
# demonstrate dequeue implementation
# using list

# Initializing a dequeue
dequeue = []

#function to display dequeue
def display():
    print(dequeue)
#function to delete element from beginning of dequeue
def deletebeg():
    temp=dequeue[0]
    dequeue.pop(0)
    print(temp,"is deleted")
#function to insert element at end of dequeue
def insertend():
    data=input("enter data to be insert-")
    dequeue.append(data)
    print("success")
#function to delete element at end of dequeue
def deleteend():
    temp=dequeue[-1]
    dequeue.pop()
    print(temp,"is deleted")
#function to insert element from beginning of dequeue
def insertbeg():
    data=input("enter data to be insert-")
    dequeue.insert(0,data)
    print("success")
#menu for dequeue operations
while(1):
    print("    menu    ")
    print("1-insert at end")
    print("2-delete in beginning")
    print("3-delete at end")
    print("4-insert in beginning")
    print("5-display")
    print("6-exit")
    choice=input("enter choice-")
```

```
if choice=="1":
    insertend()
elif choice=="2":
    deletebeg()
elif choice=="3":
    deleteend()
elif choice=="4":
    insertbeg()
elif choice=="5":
    display()
elif choice=="6":
    exit(0)
```

The output of the program is

```
   menu
1-insert at end
2-delete in beginning
3-delete at end
4-insert in beginning
5-display
6-exit
enter choice-1
enter data to be insert-5
success
   menu
1-insert at end
2-delete in beginning
3-delete at end
4-insert in beginning
5-display
6-exit
enter choice-1
enter data to be insert-6
success
   menu
1-insert at end
2-delete in beginning
3-delete at end
4-insert in beginning
5-display
6-exit
enter choice-4
enter data to be insert-3
success
```

```
    menu
1-insert at end
2-delete in beginning
3-delete at end
4-insert in beginning
5-display
6-exit
enter choice-5
['3', '5', '6']
    menu
1-insert at end
2-delete in beginning
3-delete at end
4-insert in beginning
5-display
6-exit
enter choice-2
3 is deleted
    menu
1-insert at end
2-delete in beginning
3-delete at end
4-insert in beginning
5-display
6-exit
enter choice-5
['5', '6']
```

```
    menu
1-insert at end
2-delete in beginning
3-delete at end
4-insert in beginning
5-display
6-exit
enter choice-4
enter data to be insert-7
success
    menu
1-insert at end
2-delete in beginning
3-delete at end
4-insert in beginning
5-display
6-exit
enter choice-5
['7', '5', '6']
    menu
1-insert at end
2-delete in beginning
3-delete at end
4-insert in beginning
5-display
6-exit
enter choice-3
6 is deleted
    menu
1-insert at end
2-delete in beginning
3-delete at end
4-insert in beginning
5-display
6-exit
enter choice-5
['7', '5']
```

Explanation: The above menu-driven program has the insertbeg, insertend, deleteend, deletebeg, and display functions for the double-ended queue.

- The insertbeg function adds an element in beginning to the de-queue.
- The deleteend function removes an element from the end of the de-queue.
- The insertend function adds an element to the end of the de-queue.
- The deletebeg function removes the element from the beginning of the de-queue.
- The display function prints every node of the de-queue.

5.6 APPLICATIONS OF QUEUES

- In real life, call center phone systems use queues to hold people calling them in order until a service representative is free.
- The handling of interruptions in real-time systems uses the concept of queues. The interrupts are handled in the same order as they arrive, that is, First Come, First Served.
- The round-robin technique for processor scheduling is implemented using queues.
- Queues are often used as buffers on portable CD players, MP3 players, and iPod playlists.

5.7 SUMMARY

- A queue is a linear collection of data elements in which the element inserted first will be the element taken out first (i.e., a queue is a FIFO data structure).
- A queue is a linear data structure in which the first element is inserted from one end, called the REAR end, and the deletion of the element takes place from the other end, called the FRONT end.
- The implementation of queues can be done in two ways: implementations through arrays/lists and implementations through linked lists.
- Insertion and deletion are the two basic operations that are performed on queues.
- A circular queue is a linear data structure in which the operations are performed based on a FIFO (First In, First Out) principle, and the first index comes after the last index.

- A priority queue is a queue in which elements are processed based on the assigned priority. Each element in a priority queue is assigned a special value called the priority of the element.

- When a priority queue is implemented using linked lists, then every node of the list will have three parts, that is, a data part, priority number of the element, and the address of the next element.

- A double-ended queue is a special type of data structure in which the insertion and deletion of elements are done at either end, that is, either at the front end or at the rear end of the queue.

- An input restricted de-queue is a queue in which deletion can be done at both ends, but insertion is done only at the rear end.

- An output restricted de-queue is a queue in which insertion can be done at both ends, but deletion is done only at the front end.

5.8   EXERCISES

5.8.1 Theory Questions

Q1.   What is a linear queue? Give a real-life example.

Q2.   What is a circular queue and how it is different from a linear queue?

Q3.   Define priority queues.

Q4.   Discuss various operations that can be performed on the queues.

Q5.   Define queues and in what ways a queue can be implemented. What do you understand about double-ended queues? Discuss the different types of de-queues in detail.

Q6.   Give some of the applications of queues.

Q7.   Why are queues known as First-In-First-Out structures?

Q8.   Explain the concept of a linked queue and how insertion and deletion take place in it.

5.8.2 Programming Questions

Q1.   Write a program to create a linear queue containing nine elements.

Q2.   Write an algorithm to implement a priority queue.

Q3.   Write code for insertion and deletion in a queue.

Q4. Give an algorithm for the insertion of an element in a circular queue. Write a program to implement a queue that allows for insertion and deletion at both ends.

Q5. Write an algorithm that reverses the elements of a queue.

Q6. Write an algorithm for insertion and deletion in a queue. Write the functions for insertion and deletion operations performed in a de-queue. Consider all possible cases.

Q7. Write a code for deleting an element from a circular queue.

Q8. Write a program to implement a priority queue using a linked list.

5.8.3 Multiple Choice Questions

Q1. New elements in the queue are always inserted from the

    a. Front end

    b. Middle

    c. Rear end

    d. Both (a) and (c)

Q2. A queue is a _____ data structure.

    a. FIFO

    b. LIFO

    c. FILO

    d. LILO

Q3. The overflow condition in the circular queue exists when

    a. Front = MAX − 1 and Rear = 0

    b. Front = 0 and Rear = MAX − 1

    c. Front = 0 and Rear = 0

    d. Front = MAX − 1 and Rear = MAX − 1

Q4. If the elements P, Q, R, and S are placed in a queue and are deleted one by one, in what order will they be deleted?

   a. PQRS

   b. SRQP

   c. PRQS

   d. SRQP

Q5. A data structure in which elements are inserted or deleted from the front as well as from the rear end is a

   a. Linear queue

   b. De-queue

   c. Priority Queue

   d. Circular Queue

Q6. A line outside a movie theater represents a _____.

   a. Linked List

   b. Array

   c. Queue

   d. Stack

Q7. In a queue, deletion is always done at the _____.

   a. Top end

   b. Back end

   c. Front end

   d. Rear end

Q8. In a priority queue, two elements with the same priority are processed on an FCFS basis.

   a. False

   b. True

Q9. The function that inserts the elements in a queue is called
_____.

    a. Push

    b. En-queue

    c. Pop

    d. De-queue

Q10. Which of the implementations of queues is better when the size of the queue is not known in advance?

    a. Linked List Representation

    b. Array Representation

    c. Both

    d. None of the above

## CHAPTER 6

SEARCHING AND SORTING

6.1 INTRODUCTION TO SEARCHING

Computer systems are often used to store large numbers. We require some search mechanism to retrieve a specific record from the large amounts of data stored in our computer system. Searching means to find whether a particular data item exists in an array/list or not. The process of finding a particular value in a list or an array is called *searching*. If that particular value is present in the array, then the search is said to be successful and the location of that particular value is returned by the searching process. However, if the value does not exist, then searching is said to be unsuccessful. There are many different search algorithms, but three of the popular searching techniques are as follows:

- Linear Search or Sequential Search
- Binary Search
- Interpolation Search

Here, we will discuss all these methods in detail.

6.2 LINEAR SEARCH OR SEQUENTIAL SEARCH

A linear search is also called a sequential search. This is a very simple technique used to search for a particular value in an array. A *linear search* works by comparing the value of the key being searched for with every element of the array in a linear sequence until a match is found. A search will be unsuccessful if all the data elements are read and the desired element is not found. The following are some important points:

- It is the simplest way to search an element in a list.
- It searches the data element sequentially, no matter whether the array is sorted or unsorted.

For Example – let us take an array/list of ten elements, which is declared as follows:

```
array = [87, 25, 14, 39, 74, 1, 99, 12, 30, 67]
```

The value to be searched for in the array is VALUE = 74, and then we search to find whether 74 exists in the array. If the value is present, then its position is returned. Here, the position of VAL = 74 is POS = 4 (index starting from zero), which is shown in the following figures.

Pass 1 – 87 is compared with 74. Since 87 is not equal to 74, we move to the next pass.

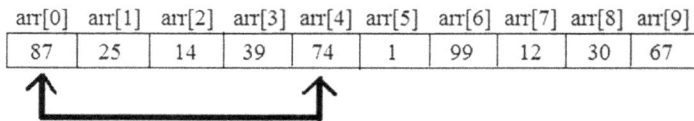

arr[0]	arr[1]	arr[2]	arr[3]	arr[4]	arr[5]	arr[6]	arr[7]	arr[8]	arr[9]
87	25	14	39	74	1	99	12	30	67

Pass 2 – 25 is compared with 74. Since 25 is not equal to 74, we move to the next pass.

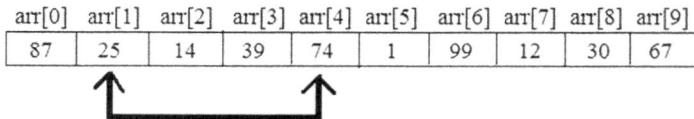

arr[0]	arr[1]	arr[2]	arr[3]	arr[4]	arr[5]	arr[6]	arr[7]	arr[8]	arr[9]
87	25	14	39	74	1	99	12	30	67

Pass 3 – 14 is compared with 74. Since 14 is not equal to 74, we move to the next pass.

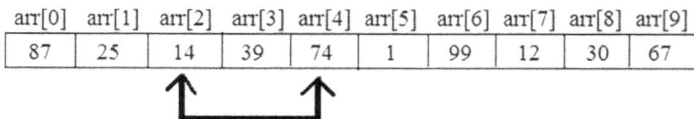

arr[0]	arr[1]	arr[2]	arr[3]	arr[4]	arr[5]	arr[6]	arr[7]	arr[8]	arr[9]
87	25	14	39	74	1	99	12	30	67

Pass 4 – 39 is compared with 74. Since 39 is not equal to 74, we move to the next pass.

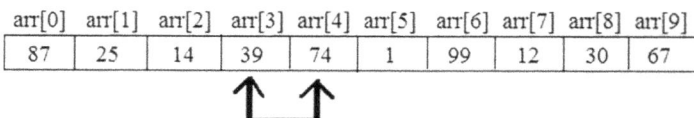

arr[0]	arr[1]	arr[2]	arr[3]	arr[4]	arr[5]	arr[6]	arr[7]	arr[8]	arr[9]
87	25	14	39	74	1	99	12	30	67

Pass 5 – 74 is compared with 74. Since 74 is equal to 74, we return the position on which 74 is present, which in this case is 4.

arr[0]	arr[1]	arr[2]	arr[3]	arr[4]	arr[5]	arr[6]	arr[7]	arr[8]	arr[9]
87	25	14	39	74	1	99	12	30	67

74 is found at POS = 4

FIGURE 6.1 Example of a linear search

In this way, a linear search is used to search for a particular value in the array. Now let us understand it further with the help of an algorithm.

> **Practical Application:**
>
> A simple real-life example of a linear search is a person who is searching for another person's contact number in a telephone directory. If the person does not know the exact name of that person but knows that the name starts with A, then she will start searching from the beginning of the telephone directory.

Algorithm for a Linear Search

Let *ARR* be an array of n elements, ARR[1], ARR[2], ARR[3], . . . ARR[n] such that VAL is the element to be searched. Then the algorithm will find the position POS of the VAL in the array ARR.

```
Step 1: START
Step 2: Set I = 0, POS = -1
Step 3: Repeat while I<N
          IF (ARR[I] = VAL)
          POS = I
      PRINT POS
              Go to Step 5
          [End of IF]
      [End of Loop]
Step 4: IF (POS = -1)
      PRINT "VALUE NOT FOUND, SEARCH UNSUCCESSFUL"
      [End of IF]
Step 5: EXIT
```

In Step 2 of the algorithm, we are initializing the values of I and POS. In Step 3, a while loop is executed in which a check is made to see whether a match is found between the current array element and VAL. If the match is found, then the position of that element is printed. In the last step, if all the elements have been compared and there is no match found, the search will be unsuccessful; that is, the value is not present in the array.

The Complexity of a Linear Search Algorithm

The execution time of a linear search is O(n), where n is the number of elements in the array. The algorithm is called a linear search because its complexity can be expressed as a linear function, which is that the number of comparisons to find the target item increases linearly with the size of the data. The best case of a linear search is when the data element to be searched for is equal to the first element of the array. The worst case occurs when the data element to be searched for is equal to the last element in the array. However, in both the cases and comparisons have to be made.

6.2.1 Drawbacks of a Linear Search

- It is a very time-consuming process, as it works sequentially.
- It can be applied only to a small amount of data.
- It is a very slow process as almost every data element is accessed in this process, especially when the data element is located near the end.

Here is a program to search an element in an array using a linear search technique.

```python
#program for a linear search
def linearsearch(arr, x):
    for i in range(len(arr)):
        if arr[i] == x:
            print("element found at index ",i+1)
#creating an array
arr=[]
size=input("enter no of elements-")
print("enter elements in array/list-")
for i in range(int(size)):
    data=input()
    arr.append(data)
x=input("enter element to search-")
linearsearch(arr,x)
```

The output of the program is

```
enter no of elements-5
enter elements in array/list-
11
23
44
16
43
enter element to search-16
element found at index   4
>>>
```

Frequently Asked Questions

Q. Explain how a linear search technique is used to search for an element.

Answer:

Suppose that ARR is an array having N elements. ITEM is the value to be searched. Then we have the following cases:

__Case 1: Unsorted List__ – The ITEM is compared with every element of the array. If the element is found, then no further comparison is required. If all the elements are compared and checked, then the ITEM is not found.

__Case 2: Sorted List__ – The ITEM is greater than the first element and smaller than the last element of the list, so the search is performed by comparing each element in the list with ITEM; otherwise, ITEM is reported as __"Not Found."__

6.3 BINARY SEARCH

A binary search is an extremely efficient search algorithm when it is compared to a linear search. A *binary search* works only when the array/list is already sorted. In a binary search, we first compare the value VAL with the data element in the middle position of the array. If the match is found, then the position POS of that element is returned; otherwise, if the value is less than that of the middle element, then we begin our search in the lower half of the array and vice versa. So, we repeat this process on the lower and upper half of the array.

6.3.1 Binary Search Algorithm

Let us now understand how this binary search algorithm works in an array.

1. Find the middle element of the array, that is, n/2 is the middle element of the array containing n elements.

2. Now, compare the middle element of the array with the data element to be searched.

    a. If the middle element is the desired element, then the search is successful.

    b. If the data element to be searched for is less than the middle element of the array, then search only the lower half of the array, that is, those elements which are on the left side of the middle element.

c. If the data element to be searched for is greater than the middle element of the array, then search only the upper half of the array, that is, those elements which are on the right side of the middle element.

Repeat these steps until a match is found.

Practical Application:

A real-life application of a binary search is when we search for a particular word in a dictionary. We first open the dictionary somewhere in the middle. Now we will compare the desired word with the first word on that page. If the desired word comes after the first word on an open page, then we look in the second half of the dictionary; otherwise, we look in the first half. Now, we again open a page in the second half and compare the desired word with the first word on that page, and the same process is repeated until we have found the desired word.

Algorithm for a Binary Search

```
Binary_Search(ARR, Lower_bound, Upper_bound, VAL)
Step 1: START
Step 2: Set BEG = lower_bound, END = upper_bound, POS = -1
Step 3: Repeat Steps 4 & 5 while BEG <= END
Step 4: Set MID = (BEG+END)/2
Step 5: IF (ARR[MID] = VAL)
        POS = MID
             PRINT POS
             Go to Step 7
        ELSE IF (ARR[MID] > VAL)
        Set END = MID - 1
        ELSE
        Set BEG = MID + 1
        [End of If]
        [End of Loop]
Step 6: IF (POS = -1)
             PRINT "VALUE NOT FOUND, SEARCH UNSUCCESSFUL"
        [End of IF]
Step 7: EXIT
```

In Step 2 of the algorithm, we initialize the values of BEG, END, and POS. In Step 3, a while loop is executed. In Step 3, the value of MID is calculated. In Step 4, we check if the value to be searched for is equal to the array value at MID. If the match is found, then the position of that element is printed. If the match is not found and the value to be searched for is less than that of the array value at MID, then the END is modified; otherwise, if the value to be searched for is greater than that of the array value at MID, then the BEG is modified. In the last step, if all the elements have been compared and there is no match found, the search has been unsuccessful; that is, the value is not present in the array.

Example:

Let us now consider an example to search for a particular value in a sorted array.

Consider an array of ten elements which is declared as follows:

```
array = [0, 10, 20, 30, 40, 50, 60, 80, 90, 100]
```

and the value to be searched for is VAL = 20. Then the algorithm proceeds as follows:

Solution –

Pass 1 –
BEG = 0, END = 10
MID = (BEG + END)/2
    = (0 + 10)/2 = 5
Now, VAL = 20 and ARR[MID] = ARR[5] = 50

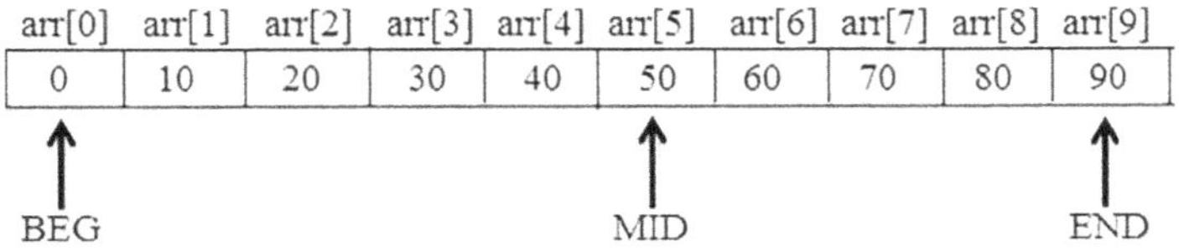

arr[0]	arr[1]	arr[2]	arr[3]	arr[4]	arr[5]	arr[6]	arr[7]	arr[8]	arr[9]
0	10	20	30	40	50	60	70	80	90

↑ BEG   ↑ MID   ↑ END

As ARR[5] = 50 > VAL = 30, therefore we now search for the value in the lower half of the array. The values of END and MID are modified, and we move to the next pass.

Pass 2 –
Now, END = MID – 1 = 4
MID = (0 + 4)/2 = 2
Now VAL = 20 and ARR[MID] = ARR[2] = 20.

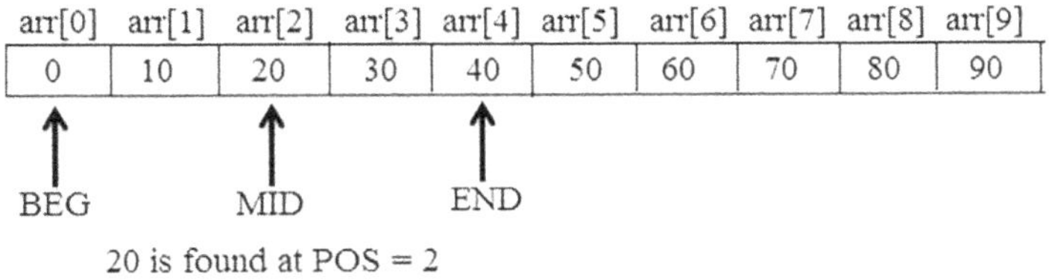

arr[0]	arr[1]	arr[2]	arr[3]	arr[4]	arr[5]	arr[6]	arr[7]	arr[8]	arr[9]
0	10	20	30	40	50	60	70	80	90

↑ BEG   ↑ MID   ↑ END

20 is found at POS = 2

FIGURE 6.2 An example of a binary search

Hence, the search is successful and VAL = 20 is found at POS = 2.

6.3.2 Complexity of a Binary Search Algorithm

In a binary search algorithm, we can see that with each comparison, the size of the search area is reduced by half. So, we can claim that the efficiency of the binary search in the worst case is $O(\log_{10}n)$, where n is the total number of elements in the array. The best case happens when the value to be searched for is equal to the value of the array in the middle.

6.3.3 Drawbacks of a Binary Search

- A binary search requires that the data elements in the array be sorted; otherwise, a binary search will not work.

- A binary search cannot be used where there are many insertions and deletions of data elements in the array.

Here is a program to search for an element in an array using the binary search technique.

```
# program for binary search
def binarysearch(arr, low, high, x):

    # Check base case
    if high >= low:

        mid = (high + low) // 2

        # If element is present at the middle itself
        if arr[mid] == x:
            return mid

        # If element is smaller than mid, then it can only
        # be present in left sub-array
        elif arr[mid] > x:
            return binarysearch(arr, low, mid - 1, x)

        # Else the element can only be present in right sub-array
        else:
            return binarysearch(arr, mid + 1, high, x)

    else:
        # Element is not present in the array
        return -1
arr=[]
size=input("enter no of elements-")
print("enter elements in array/list-")
for i in range(int(size)):
    data=input()
    arr.append(data)
x=input("enter element to search-")
result=binarysearch(arr,0,len(arr)-1,x)
```

```
if result != -1:
    print("Element is present at index", str(result+1))
else:
    print("Element is not present in array")
```

The output of the program is as follows:

```
enter no of elements-5
enter elements in array/list-
11
22
33
44
55
enter element to search-11
Element is present at index 1
>>> |
```

Frequently Asked Questions

Q. What is a binary search? Explain.

Answer:

A binary search is a search technique that is used to find an element in an array. It works very efficiently with a sorted list. In a binary search, the element to be searched for is compared with the middle element of the array. If the value to be searched for is less than the middle element, we search in the lower half of the array and vice versa.

6.4 INTERPOLATION SEARCH

An *interpolation search*, also known as an extrapolation search, is a technique for searching for a particular value in an ordered array. This search technique is more efficient than a binary search if the elements in the array are sorted. The technique of an interpolation search is similar to when we are searching for "Andersen" in the telephone directory; we don't start in the middle, because we know that it will be near the extreme left, so we start from the front and work from there. That is the main idea of an interpolation search; that is, instead of dividing the list into fixed halves, we cut it by an amount that seems most likely to succeed.

> **Practical Application:**
>
> If we want to search for "Adams" in the directory, then we always search in the extreme left of the directory.

6.4.1 The Interpolation Search Algorithm

In each step of this searching technique, the remaining search area for the value to be searched for is calculated. The calculations are done on the values at the bounds of the search area and the value which is to be searched. Therefore, the value found at this position is compared with the value to be searched. If both values are equal, then the search is said to be successful. If both values are unequal, then depending upon the comparison done, the remaining search area is reduced to the part just before or after the initial position.

Consider an array ARR of n elements in which the elements are arranged in a sorted manner. Initially low is set to 0 and high is set to n-1. Now we are searching a value VAL in ARR between ARR[LOW] and ARR[HIGH]. Then, in this case, MID will be calculated by the following formula:

> **MID** = LOW + (HIGH – LOW) X ((VAL – ARR[LOW] / ARR[HIGH] – ARR[LOW]))

If the value VAL is found at MID, then the search is complete; otherwise, if the value is lower than ARR[MID], reset HIGH = MID – 1, and if the value is greater than ARR[MID], reset LOW = MID + 1. Repeat these steps until the value is found.

Hence, we can say that the interpolation search is very similar to the binary search technique. The main difference between the techniques is that in a binary search the value selected is always the middle value of the list, and it discards half the values based on the comparison between the value to be searched for and the value found at the estimated position. Let us understand the interpolation search with the help of an algorithm.

Algorithm for an Interpolation Search

```
INTERPOLATION_SEARCH(ARR,    Lower_bound,    Upper_bound,
VAL)
Step 1: START
Step 2: Set LOW = lower_bound, HIGH = upper_bound, POS = -1
Step 3: Repeat Steps 4 & 5 while LOW<= HIGH
```

```
Step 4: Set MID = LOW + (HIGH - LOW) X ((VAL - ARR[LOW] /
ARR[HIGH] - ARR[LOW]))
Step 5: IF (ARR[MID] = VAL)
        POS = MID
        PRINT POS
            Go to Step 7
        ELSE IF (ARR[MID] > VAL)
        Set HIGH = MID - 1
        ELSE
        Set LOW = MID + 1
        [End of If]
        [End of Loop]
Step 6: IF (POS = -1)
        PRINT "VALUE NOT FOUND, SEARCH UNSUCCESSFUL"
        [End of IF]
Step 7: EXIT
```

Example: Consider an array of seven numbers which is declared as

`array = [5, 16, 23, 34, 45, 56, 65]`

and the value to be searched for is 45.

Solution –

Pass 1 –

LOW = 0, HIGH = 7 – 1= 6, VAL = 45

ARR[LOW] = ARR[0] = 5, ARR[HIGH] = ARR[6] = 65

Now MID = LOW + (HIGH – LOW) × ((VAL – ARR[LOW]) / (ARR[HIGH] – ARR[LOW]))

$$= 0 + (6 - 0) \times ((45 - 5) / (65 - 5))$$
$$= 0 + 6 \times (40 / 60) = 4$$

If(VAL == ARR[MID]) i.e. 45 == ARR[4] = 45, 45 = 45

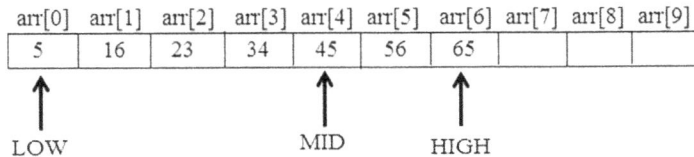

45 is found at POS = 4

FIGURE 6.3 An interpolation search

Hence, a value is found.

6.4.2 Complexity of the Interpolation Search Algorithm

The interpolation search makes about $\log_{10}(\log_{10} n)$ comparisons when there are n elements in the list and the elements are uniformly distributed. The worst case happens when the number of elements increases exponentially; in that case, the algorithm can take up to $O(n)$ comparisons.

Here is a program to search for an element in an array using the interpolation search technique.

```python
# Python program to implement interpolation search

# If x is present in arr[0..n-1], then returns
# index of it, else returns -1
def interpolationSearch(arr, n, x):
    # Find indexes of two corners
    lo = 0
    hi = (n - 1)

    # Since array is sorted, an element present
    # in array must be in range defined by corner
    while lo <= hi and x >= arr[lo] and x <= arr[hi]:
        if lo == hi:
            if arr[lo] == x:
                return lo;
            return -1;

        # Probing the position with keeping
        # uniform distribution in mind.
        pos  = lo + int(((float(hi - lo) /
            ( arr[hi] - arr[lo])) * ( x - arr[lo])))

        # Condition of target found
        if arr[pos] == x:
            return pos

        # If x is larger, x is in upper part
        if arr[pos] < x:
            lo = pos + 1;

        # If x is smaller, x is in lower part
        else:
            hi = pos - 1;

    return -1
arr=[]
size=input("enter no of elements-")
print("enter elements in array/list-")
for i in range(int(size)):
    data=int(input())
    arr.append(data)
```

```
x=int(input("enter element to search-"))
n = len(arr)

index = interpolationSearch(arr, n, x)

if index != -1:
    print ("Element found at index",index+1 )
else:
    print ("Element not found")
```

The output of the program is as follows:

```
enter no of elements-5
enter elements in array/list-
11
22
33
44
55
enter element to search-33
Element found at index 3
>>>
```

6.5   INTRODUCTION TO SORTING

Sorting refers to the process of arranging the data elements of an array in a specified order, that is, either in ascending or descending order. For example, it is practically impossible for us to find a name in the telephone directory if the names in it are not in alphabetical order. However, the same can be true for dictionaries, book indexes, and bank accounts. Hence, the convenience of having sorted data is unquestionable. Retrieval of information becomes much easier when the data is stored in some specified order. Therefore, sorting is a very important application in computer science.

Let us take an array that is declared and initialized as

```
array = [10, 25, 17, 8, 30, 3]
```

Then, the array after applying the sorting technique is

```
array = [3, 8, 10, 17, 25, 30]
```

A sorting algorithm can be defined as an algorithm that puts the data elements of an array/ list in a certain order, that is, either numerical order or any predefined order. Many sorting algorithms are available and are widely used according to the different environments required by the different sorting methods.

The two basic categories of sorting methods are

1. **Internal Sorting** – This refers to the sorting of the data elements stored in the computer's main memory.

2. **External Sorting** – This refers to the sorting of the data elements stored in the files. It is applied when the amount of data is large and cannot be stored in the main memory.

6.5.1 Types of Sorting Methods

The various sorting methods are

1. Selection Sort

2. Insertion Sort

3. Merge Sort

4. Bubble Sort

5. Quicksort

Let us discuss all of them in detail.

1. Selection Sort

Selection sort is a sorting technique that works by finding the smallest value in the array and placing it in the first position. After that, it then finds the second smallest value and places it in the second position. This process is repeated until the whole array is sorted. Thus, the selection sort works by finding the smallest unsorted element remaining in the entire array and then swapping it with the element in the next position to be filled. It is a very simple technique, and it is also easier to implement than other sorting techniques. The selection sort is used for sorting files with large records.

Selection Sort Technique

Let us take an array ARR with N elements in it. Now, the selection sort technique works as follows:

First of all, we find the smallest value in the entire array, and we place that value in the first position of the array. Then, we find the second smallest value in the array, and we place it in the second position of the array. Now, we repeat this process until the whole array is sorted.

Pass 1 – Find the position POS of the smallest value in the array of N elements and interchange ARR[POS] with ARR[0]. Hence, ARR[0] is sorted.

Pass 2 – Find the position POS of the smallest value in the array of N-1 elements and interchange ARR[POS] with A[1]. Hence, A[1] is sorted.

.

.

.

Pass N–1– Find the position POS of the smaller of the elements of ARR[N–2] and ARR[N–1] and interchange ARR[POS] with ARR[N–2]. Hence, ARR[0], ARR[1], . . . ARR[N–1] is sorted.

Let us discuss this concept with the help of a detailed algorithm.

Algorithm for a Selection Sort

Consider an array ARR having N elements from ARR[0] to ARR[N–1]. I and J are the looping variables, and POS is the swapping variable.

```
SELECTION SORT(ARR, N)
Step 1: START
Step 2: Repeat Steps 3 & 4 for I = 1 to N - 1
Step 3: Call MIN(ARR, I, N, POS)
Step 4: Swap ARR[I] with ARR[POS]
[End of Loop]
Step 5: EXIT

MIN(ARR, I, N, POS)
Step 1:Set SMALLEST = ARR[I]
Step 2:Set POS = I
Step 3: Repeat Step 4 for J = I + 1 to N - 1
Step 4: IF (ARR[J] < SMALLEST)
           Set SMALLEST = ARR[J]
Set POS = J
[End of IF]
[End of Loop]
Step 5: Return POS
```

Example – Sort the given array using the selection sort.

arr[0]	arr[1]	arr[2]	arr[3]	arr[4]
4	14	29	11	35

Solution:

Pass	POS	Array[0]	Array[1]	Array[2]	Array[3]	Array[4]
1	4	4	14	29	11	35
2	3	4	11	29	14	35
3	3	4	11	14	29	35
4	3	4	11	14	29	35
5	4	4	11	14	29	35

Hence, after sorting the new array is

arr[0]	arr[1]	arr[2]	arr[3]	arr[4]
4	11	14	29	35

FIGURE 6.4 An example of a selection sort

The Complexity of the Selection Sort Algorithm

The selection sort is the simplest sorting technique. In this method, if there are n elements in the array, then (n–1) comparisons or iterations are made. Thus, the selection sort technique has a complexity of $O(n^2)$.

Here is a program to sort an array using the selection sort method.

```
# Python program for implementation of Selection
# Sort
A = []
size=int(input("enter size of array-"))
print("enter elements of array to be sorted-")
for i in range(size):
    data=int(input())
    A.append(data)

# Traverse through all array elements
for i in range(len(A)):

    # Find the minimum element in remaining
    # unsorted array
    min_ = i
    for j in range(i+1, len(A)):
        if A[min_] > A[j]:
            min_ = j

    # Swap the found minimum element with
    # the first element
    A[i], A[min_] = A[min_], A[i]
```

```
# Driver code to test above
print ("Sorted array")
for i in range(len(A)):
    print("%d" %A[i])
```

The output of the program is as follows:

```
enter size of array-5
enter elements of array to be sorted-
4
2
8
9
1
Sorted array
1
2
4
8
9
```

Frequently Asked Questions

Q. Define the selection sort technique.

Answer:

The selection sort is a sorting technique that works by finding the smallest element from the array and placing it in the first position. It then finds the second smallest element and places it in the second position. Hence, this procedure is repeated until the whole array is sorted.

2. Insertion Sort

The insertion sort is another very simple sorting algorithm that works just like its name suggests; that is, it inserts each element into its proper position in the concluding list. To limit the waste of memory or, we can say, to save memory, most implementations of an insertion sort work by moving the current element past the already sorted elements and repeatedly swapping or interchanging it with the preceding element until it is placed in its correct position.

Practical Application:

This technique is used when ordering a deck of cards in the card game bridge.

Insertion Sort Technique

Pass 1 – Initially, there is only one element in the list which is already sorted. Hence, we proceed to the next steps.

Pass 2 – During the first iteration, the first and the second element of the list are compared. The smaller value occupies the first position of the list.

Pass 3 – During the second iteration, the first three elements of the list are compared. The smaller value occupies the first position in the list. The second position is occupied by the second smallest element, and so on.

.
.
.

This procedure is repeated for all the elements of the array up to (n−1) iterations.

Algorithm for an Insertion Sort

```
INSERTION SORT(ARR, N)

Step 1: START
Step 2: Repeat Steps 3 to 6 for I = 1 to N - 1
Step 3: Set POS = ARR[I]
Step 4: Set J = I - 1
Step 5: Repeat while POS <= ARR[J]
            Set ARR[J + 1] = ARR[J]
            Set J = J - 1
[End of Inner while loop]
Step 6: Set ARR[J + 1] = POS
[End of Loop]
Step 7: EXIT
```

In the previous algorithm, in Step 2, a for loop is executed which is repeated for every element in the array. In Step 3, we store the value of the Ith element in POS. In Step 5, again a loop is executed in which the new elements after sorting are placed. At last, the element is stored at the $(J+1)^{th}$ position.

For Example – Consider the following array. Sort the given values in the array using the insertion sort technique.

arr[0]	arr[1]	arr[2]	arr[3]	arr[4]	arr[5]
39	54	10	28	95	7

FIGURE 6.5 An example of an insertion sort

Solution – ⬚ sorted   ⬚ unsorted

Pass 1 – Initially, ARR[0] is sorted. Move to the next pass.

39	54	10	28	95	7

Pass 2 – Now 39 and 54 are compared. 39 < 54, so ARR[0] = 39 and ARR[1] = 54.

39	54	10	28	95	7

Pass 3 – 39, 54, and 10 are compared. 10 < 39 and 54, so ARR[0] = 10, now 39 < 54, hence ARR[1] = 39 and ARR[2] = 54.

10	39	54	28	95	7

Pass 4 – As 28 < 39 and 54, so ARR[1] = 28.

10	28	39	54	95	7

Pass 5 – In this case, 95 is greater than all the values, so there is no need for swapping.

10	28	39	54	95	7

Pass 6 – 7 is the smallest value, so ARR[0] = 7.

Therefore, after sorting the new array is

7	10	28	39	54	95

The Complexity of an Insertion Sort

In an insertion sort, the best case happens when the array is already sorted, and in that case, the running time of the algorithm is $O(n)$(i.e., linear running time). The worst case happens when the array is sorted in the reverse order. Thus, in that case, the running time of the algorithm is $O(n^2)$ (i.e., the quadratic running time).

Here is a program to sort an array using the insertion sort method.

```
# Python program for implementation of Insertion Sort

# Function to do insertion sort
def insertionSort(arr):

    # Traverse through 1 to len(arr)
    for i in range(1, len(arr)):

        key = arr[i]

        # Move elements of arr[0..i-1], that are
        # greater than key, to one position ahead
        # of their current position
        j = i-1
        while j >=0 and key < arr[j] :
                arr[j+1] = arr[j]
                j -= 1
        arr[j+1] = key
arr = []
size=int(input("enter size of array-"))
print("enter elements of array to be sorted-")
for i in range(size):
    data=int(input())
    arr.append(data)
# Driver code to test above
insertionSort(arr)
print ("Sorted array is:")
for i in range(len(arr)):
    print ("%d" %arr[i])
```

The output of the program is as follows:

```
enter size of array-5
enter elements of array to be sorted-
7
2
9
3
1
Sorted array is:
1
2
3
7
9
```

3. Merge Sort

The *merge sort* is a sorting method that follows the divide and conquers approach. The divide and conquer approach is a very good approach in which divide refers to partitioning the array having n elements into two sub-arrays of n/2 elements each. However, if there are no elements present in the list/array or if an array contains only one element, then it is already sorted. However, if an array has more elements, then it is divided into two sub-arrays containing equal elements in them. Conquer is the process of sorting the two sub-arrays recursively using the merge sort. Finally, the two sub-arrays are merged into one single sorted array.

Merge Sort Techniques

1. If the array has zero or one element in it, then there is no need to sort that array as it is already sorted.

2. Otherwise, if there are more elements in the array, then divide the array into two sub-arrays containing equal elements.

3. Each sub-array is now sorted recursively using the merge sort.

4. Finally, the two sub-arrays are merged into a single sorted array.

Algorithm of a Merge Sort

```
MERGE SORT(ARR, BEG, END)

Step 1: START
Step 2: IF (BEG < END)
Step 3: Set MID = (BEG + END)/2
        Call MERGE SORT (ARR, BEG, ENDMID )
        Call MERGE SORT (ARR, MID + 1, END )
        Call MERGE (ARR, BEG, MID, END)
        [End ofIf]
Step 4: EXIT

MERGE(ARR, BEG, MID, END)

Step 1: START
Step 2: Set I = BEG, J = MID + 1, K = 0
Step 3: Repeat while (I <= MID)   && (J <= END)
IF (ARR[I] > ARR[J])
            Set TEMP[K] = ARR[J]
            Set J = J + 1
            Set K = K + 1
ELSE IF (ARR[J] > ARR[I])
            Set TEMP[K] = ARR[I]
            Set I = I + 1
            Set K = K + 1
ELSE
            Set TEMP[K] = ARR[J]
            Set J = J + 1
            Set K = K + 1
            Set TEMP[K] = ARR[I]
```

```
                        Set I = I + 1
                        Set K = K + 1
[End of If]
[End of Loop]

Step 4: (Copying the remaining elements of left sub array if
any)
Repeat while (I <= MID)
                Set TEMP[K] = ARR[I]
                Set I = I + 1
                Set K = K + 1
        [End of Loop]

Step 5: (Copying the remaining elements of right sub array if
any)
Repeat while (J <= END)
                Set TEMP[K] = ARR[J]
                Set I = I + 1
                Set K = K + 1
[End of Loop]

Step 6: Set IND = 0

Step 7: Repeat while (IND < K)
Set ARR[IND] = ARR[IND]
                Set IND = IND + 1
[End of Loop]

Step 8: EXIT
```

For Example – Sort the following array using merge sort.

```
array = [ 40, 10, 86, 44, 93, 26, 69, 17 ]
```

Solution –

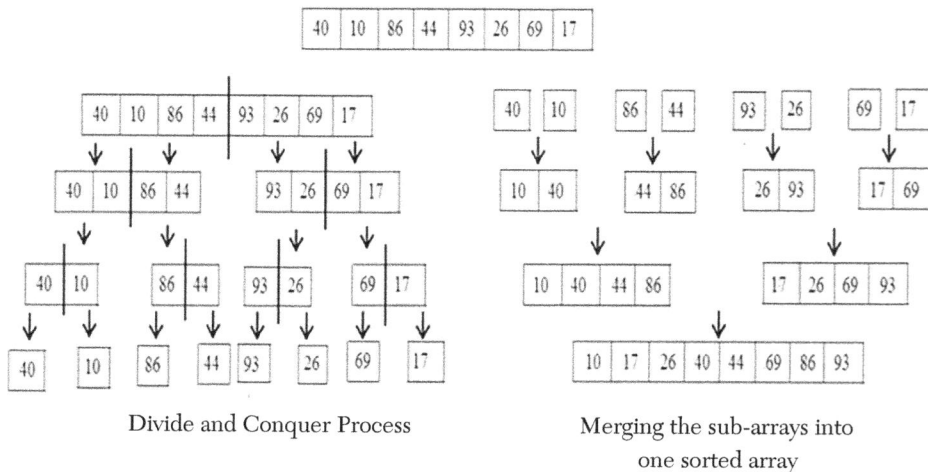

Divide and Conquer Process   Merging the sub-arrays into
                                            one sorted array

FIGURE 6.6 An example of a merge sort

From the previous example, we can see how the merge sort algorithm works. First, the merge sort algorithm recursively divides the array into smaller sub-arrays. After dividing the array into smaller parts, we call the function Merge() to merge all the sub-arrays to form a single sorted array.

The Complexity of a Merge Sort

The running time of the merge sort algorithm is $O(n \log_{10} n)$. This runtime remains the same in the average as well as in the worst case of the merge sort algorithm. Although it has an optimal time complexity, sometimes this runtime can be $O(n)$.

Here is a program to sort an array using the merge sort method.

```python
# Python program for implementation of a Merge Sort

# Merges two subarrays of arr[].
# First subarray is arr[l..m]
# Second subarray is arr[m+1..r]
def merge(arr, l, m, r):
    n1 = m - l + 1
    n2 = r- m

    # create temp arrays
    L = [0] * (n1)
    R = [0] * (n2)
# Copy data to temp arrays L[] and R[]
for i in range(0 , n1):
    L[i] = arr[l + i]

for j in range(0 , n2):
    R[j] = arr[m + 1 + j]

# Merge the temp arrays back into arr[l..r]
i = 0     # Initial index of first subarray
j = 0     # Initial index of second subarray
k = l     # Initial index of merged subarray

while i < n1 and j < n2 :
    if L[i] <= R[j]:
        arr[k] = L[i]
        i += 1

    else:
        arr[k] = R[j]
        j += 1
    k += 1

# Copy the remaining elements of L[], if there
# are any
while i < n1:
    arr[k] = L[i]
    i += 1
    k += 1
```

```
# Copy the remaining elements of R[], if there
# are any
while j < n2:
    arr[k] = R[j]
    j += 1
    k += 1

# l is for left index and r is right index of the
# sub-array of arr to be sorted
def mergeSort(arr,l,r):
    if l < r:

        # Same as (l+r)//2, but avoids overflow for
        # large l and h
        m = (l+(r-1))//2

        # Sort first and second halves
        mergeSort(arr, l, m)
        mergeSort(arr, m+1, r)
        merge(arr, l, m, r)

# Driver code to test above
arr = []
size=int(input("enter size of array-"))
print("enter elements of array to be sorted-")
for i in range(size):
    data=int(input())
    arr.append(data)
n = len(arr)

mergeSort(arr,0,n-1)
print ("\n\nSorted array is")
for i in range(n):
    print ("%d" %arr[i])
```

The output of the program is as follows:

```
enter size of array-5
enter elements of array to be sorted-
6
3
9
2
4


Sorted array is
2
3
4
6
9
```

4. Bubble Sort

The bubble sort, also known as an exchange sort, is a very simple sorting method. It works by repeatedly moving the largest element to the highest position of the array. In the bubble sort, we compare two elements at a time, and swapping is done if they are wrongly placed. If the element at a lower index or position is greater than the element at a higher index, then both the elements are interchanged so that the smaller element is placed before the bigger one. This process is repeated until the list becomes sorted. The bubble sort gets its name from the way that the smaller elements "bubble" to the top of the array. This sorting technique only uses comparisons to operate on the elements. Hence, we can also call it a comparison sort.

Bubble Sort Technique

The basic idea applied for a bubble sort is to let us assume if an array ARR contains n elements, then the number of iterations required to sort the array will be (n – 1).

Pass 1 – During the first iteration, the largest value in the array is placed at the last position.

Pass 2 – During the second iteration, the second largest value of the array is placed in the second last position.

Pass 3 – During the third iteration, the third-largest value of the array is placed in the third last position and so on.

This procedure is repeated until all the elements in the array are scanned and are placed in their correct position, which means that the array is sorted.

Algorithm of a Bubble Sort

```
BUBBLE SORT(ARR, N)

Step 1: START
Step 2: Repeat Step 3 for I = 0 to N - 1
Step 3: Repeat for J = 0 to N - 1
Step 4: IF (ARR[J] > ARR[J+1])
             INTERCHANGE ARR[J] & ARR[J + 1]
[End of Inner Loop]
[End of Outer Loop]
Step 5: EXIT
```

For Example – Consider the following array. Sort the given values in the array using the bubble sort technique.

arr[0]	arr[1]	arr[2]	arr[3]	arr[4]
40	50	20	90	30

FIGURE 6.7 An example of the bubble sort

Solution – In the given array, the number of elements in the array is 5, so the number of iterations will be $(n - 1) = 4$.

Pass 1 –

40	50	20	90	30

    a. 40 and 50 are compared. Since 40 < 50, no swapping is done.

40	50	20	90	30

    b. 50 and 20 are compared. Since 50 > 20, swapping is done.

40	20	50	90	30

    c. 50 and 90 are compared. Since 50 < 90, no swapping is done.

40	20	50	90	30

    d. 90 and 30 are compared. Since 90 > 30, swapping is done.

40	20	50	30	90

At the end of the first pass, the largest element in the array is placed at the highest position in the array, but all the other elements are still unsorted. Let us now proceed to Pass 2.

Pass 2 –

40	20	50	30	90

    a. 40 and 20 are compared. Since 40 > 20, swapping is done.

20	40	50	30	90

    b. 40 and 50 are compared. Since 40 < 50, no swapping is done.

20	40	50	30	90

   c. 50 and 30 are compared. Since 50 > 30, swapping is done.

20	40	30	50	90

At the end of the second pass, the second largest element in the array is placed at the second last position in the array, but all the other elements are still unsorted. Let us now proceed to Pass 3.

Pass 3 –

20	40	30	50	90

   a. 20 and 40 are compared. Since 20 < 40, no swapping is done.

20	40	30	50	90

   b. 40 and 30 are compared. Since 40 > 30, swapping is done.

20	30	40	50	90

At the end of the third pass, the third largest element in the array is placed at the third-largest position in the array, but all the other elements are still unsorted. Let us now proceed to Pass 4.

Pass 4 –

20	40	30	50	90

   a. 20 and 40 are compared. Since 20 < 40, no swapping is done.

At the end of the fourth pass, we can see that all the elements in the list are sorted. Hence, after sorting, the new array is

20	40	30	50	90

The Complexity of the Bubble Sort

In the best case, the running time of the bubble sort is $O(n)$, that is, when the array is already sorted. Otherwise, its level of complexity in average and worst cases is $O(n^2)$.

Here is a program to sort an array using the bubble sort method.

```
# Python program for implementation of a bubble sort

def bubbleSort(arr):
    n = len(arr)

    # Traverse through all array elements
    for i in range(n-1):
    # range(n) also works but the outer loop will repeat one
time more than needed.

            # Last i elements are already in place
            for j in range(0, n-i-1):
                # traverse the array from 0 to n-i-1
                # Swap if the element found is greater
                # than the next element
                if arr[j] > arr[j+1] :
                    arr[j], arr[j+1] = arr[j+1], arr[j]
# Driver code to test above
arr = []
size=int(input("enter size of array-"))
print("enter elements of array to be sorted-")
for i in range(size):
    data=int(input())
    arr.append(data)

bubbleSort(arr)

print ("Sorted array is:")
for i in range(len(arr)):
    print ("%d" %arr[i]),
```

The output of the program is as follows:

```
enter size of array-5
enter elements of array to be sorted-
7
3
9
1
6
Sorted array is:
1
3
6
7
9
```

5. Quicksort

Quicksort, also known as partition exchange sort, was developed by C. A. R. Hoare. It is a widely used sorting algorithm that also uses the divide and conquers approach as we have discussed in merge sort. Here also, we divide a single unsorted array into its two smaller sub-arrays. The divide and conquer method involves dividing the bigger problem into two smaller problems, and then those two smaller problems into smaller problems, and so on. Like a merge sort, if there are no elements in the array or if an array contains only one element, then it is already sorted. A Quicksort algorithm is faster than all the other sorting algorithms which have the time complexity $O(n \log_{10} n)$.

How Quicksort Works

1. An element called *pivot* is selected from the array elements.

2. After choosing the pivot element, all the elements of the array are rearranged such that all the elements less than the pivot element will be on the left side, and all the elements greater than the pivot element will be placed on the right side of the pivot element. After rearranging all the elements, the pivot is now placed in its final position. Thus, this process is known as partitioning.

3. Now, the two sub-arrays obtained will be recursively sorted.

The Quicksort Technique

1. Initially set the index of the first element to LEFT and POS. Similarly, set the index of the last element to RIGHT. Now, LEFT = 0, POS = 0, RIGHT = N – 1 (assuming n elements in the array).

2. We start with the last element, which is pointed to by RIGHT, and we traverse each element in the array from right to left, comparing each element with the first element pointed to by POS. ARR[POS] should always be less than ARR[RIGHT].

   – If ARR[POS] is less than ARR[RIGHT], then continue comparing until RIGHT = POS. If RIGHT = POS then it means that pivot is placed in its correct position.

   – If ARR[RIGHT] < ARR[POS], then swap the two values and go to the next step.

   – Set POS = RIGHT.

3. We start from the first element, which is pointed to by LEFT, and we traverse every element in the array from left to right, comparing each element with the element pointed to by POS. ARR[POS] should always be greater than ARR[LEFT].

   – If ARR[POS] is greater than ARR[RIGHT], then continue comparing until LEFT = POS. If LEFT = POS then it means that pivot is placed in its correct position.

— If ARR[LEFT] > ARR[POS], then swap the two values and go to the previous step.

— Set POS = LEFT.

Algorithm for the Quicksort

```
QUICK SORT(ARR, BEG, END)
Step 1: START
Step 2: IF (BEG < END)
          Call PARTITION (ARR, BEG, END, POS)
          Call QUICK SORT (ARR, BEG, POS - 1)
Call QUICK (ARR, POS + 1, END)
     [End of If]
Step 3: EXIT

PARTITION(ARR, BEG, END, POS)
Step 1: START
Step 2: Set LEFT = BEG, RIGHT = END, POS = BEG, TEMP = 0
Step 3: Repeat Steps 4 to 7 while TEMP = 0
Step 4: Repeat while ARR[RIGHT] >= ARR[POS]&& POS != RIGHT
              Set RIGHT = RIGHT - 1
[End of Loop]
Step 5: IF (POS = RIGHT)
              Set TEMP = 1
ELSE IF (ARR[POS] > ARR[RIGHT])
              INTERCHANGE ARR[POS] with ARR[RIGHT]
              Set POS = RIGHT
[End of If]
Step 6: IF TEMP = 0
              Repeat while ARR[POS] >= ARR[LEFT] && POS != LEFT
              Set LEFT = LEFT + 1
[End of Loop]
Step 7: IF (POS = LEFT)
              Set TEMP = 1
ELSE IF (ARR[LEFT] > ARR[POS])
              INTERCHANGE ARR[POS] with ARR[LEFT]
              Set POS = LEFT
[End of If]
[End of If]
[End of Loop]
Step 8: EXIT
```

For Example – Sort the values given in the following array using the Quicksort algorithm.

arr[0]	arr[1]	arr[2]	arr[3]	arr[4]	arr[5]
25	7	39	17	30	52

FIGURE 6.8 An example of a Quicksort

Solution

Step 1 – The first element is chosen as the pivot. Now, set POS = 0, LEFT = 0, RIGHT = 5.

25	7	39	17	30	52

POS, LEFT RIGHT

Step 2 – Traverse the list from right to left. Since ARR[POS] < ARR[RIGHT], that is, 25 < 52, RIGHT = RIGHT – 1 = 4.

25	7	39	17	30	52

POS, LEFT RIGHT

Step 3 – Since ARR[POS] < ARR[RIGHT], that is, 25 < 30, RIGHT = RIGHT – 1 = 3.

25	7	39	17	30	52

POS, LEFT RIGHT

Step 4 – Since ARR[POS] > ARR[RIGHT], that is, 25 < 17, we swap the two values and set POS = RIGHT.

17	7	39	25	30	52

LEFTRIGHT, POS

Step 5 – Traverse the list from left to right. Since ARR[POS] > ARR[LEFT], that is, 25 > 17, LEFT = LEFT + 1.

17	7	39	25	30	52

LEFT RIGHT, POS

Step 6 – Since ARR[POS] > ARR[LEFT], that is, 25 > 7, LEFT = LEFT + 1.

17	7	39	25	30	52

LEFT RIGHT, POS

Step 7 – Since ARR[POS] < ARR[LEFT], that is, 25 < 39, we swap the values and set POS = LEFT.

17	7	25	39	30	52

LEFT, POS RIGHT

Step 8 – Traverse the list from right to left. Since ARR[POS] < ARR[LEFT], RIGHT = RIGHT - 1.

17	7	25	39	30	52

LEFT, POS, RIGHT

Now, RIGHT = POS, so now the process is over and the pivot element of the array, that is, 25, is placed in its correct position. Therefore, all the elements that are smaller than 25 are placed before it and all the elements greater than 25 are placed after it. Hence, 17 and 7 are the elements in the left sub-array, and 39, 30, and 52 are the elements in the right sub-array, which are both sorted.

The Complexity of Quicksort

The running time efficiency of a Quicksort is $O(n \log_{10} n)$ in the average and the best case. However, the worst case happens if the array is already sorted and the leftmost element is selected as the pivot element. In the worst case, its efficiency is $O(n^2)$.

Here is a program to sort an array using the Quicksort method.

```python
# Python program for the implementation of the Quicksort method

# This function takes last element as pivot, places
# the pivot element at its correct position in a sorted
# array, and places all smaller (smaller than pivot)
# to the left of pivot and all greater elements to the right
# of pivot
def partition(arr,low,high):
    i = ( low-1 )          # index of smaller element
    pivot = arr[high]      # pivot

    for j in range(low , high):

        # If current element is smaller than or
        # equal to pivot
        if   arr[j] <= pivot:

            # increment index of smaller element
            i = i+1
            arr[i],arr[j] = arr[j],arr[i]
```

```
        arr[i+1],arr[high] = arr[high],arr[i+1]
        return ( i+1 )
# The main function that implements Quicksort
# arr[] --> Array to be sorted,
# low   --> Starting index,
# high  --> Ending index

# Function to do Quicksort
def quickSort(arr,low,high):
    if low < high:

        # pi is partitioning index, arr[p] is now
        # at right place
        pi = partition(arr,low,high)

        # Separately sort elements before
        # partition and after partition
        quickSort(arr, low, pi-1)
        quickSort(arr, pi+1, high)

# Driver code to test above
arr = []
size=int(input("enter size of array-"))
print("enter elements of array to be sorted-")
for i in range(size):
    data=int(input())
    arr.append(data)
n = len(arr)
quickSort(arr,0,n-1)
print ("Sorted array is:")
for i in range(n):
    print ("%d" %arr[i])
```

The output of the program is as follows:

```
enter size of array-5
enter elements of array to be sorted-
9
4
6
1
3
Sorted array is:
1
3
4
6
9
```

6.6 EXTERNAL SORTING

External sorting is a sorting technique that is used when the amount of data is massive. When a large amount of data has to be sorted, it is not possible to bring it into the main memory (RAM). Therefore, a secondary memory needs to be used. Also, at the same time, some portion of data is brought into the main memory from the secondary memory for sorting based on the availability of the storage space of the main memory. After the data is sorted, it is sent back to the secondary memory. Now, the next portion of the data is brought into the main memory, and after sorting it is sent back to the secondary memory. This procedure is repeated until all the data is sorted. Here, each portion is called a *segment*. The time required for sorting is greater because time will be spent transferring the data from secondary memory to the main memory. The merge sort algorithm is widely and commonly used in external sorting, which has already been discussed.

External sorting is used in database applications for performing different kinds of operations like join, union, and projection. It is also used to update a master file from a transaction file (for example, if we are updating a company file based on the new employees, existing employees, and locations). Duplicate records or data can also be removed from external sorting.

6.7 SUMMARY

- The process of finding a particular value in a list or an array is called searching. If that particular value is present in the array, then the search is said to be successful, and the location of that particular value is retrieved by the search process.

- The linear search, binary search, and interpolation search are the most commonly used searching techniques.

- Linear search works by comparing the values to be searched for with every element of the array in a linear sequence until a match is found.

- Binary search works efficiently when the list is sorted. In a binary search, we first compare the value VAL with the data element in the middle position of the array.

- Interpolation search, also known as extrapolation search, is a technique for searching for a particular value in an ordered array. In each step of this searching technique, the remaining search area for the value to be searched for is calculated. The calculations are done on the values at the bounds of the search area and the value which is to be searched.

- Sorting refers to the technique of arranging the data elements of an array in a specified order, that is, either in ascending or descending order.
- Selection sort is a sorting technique that works by finding the smallest value in the array and placing it in the first position. After that, it then finds the second smallest value and places it in the second position. This process is repeated until the whole array is sorted.
- The insertion sort works by moving the current data element past the already sorted data elements and repeatedly interchanging it with the preceding element until it is in the correct place.
- The merge sort is a sorting method that follows the divide and conquer approach. Divide means partitioning the array having n elements into two sub-arrays of n/2 elements each. Conquer is the process of sorting the two sub-arrays recursively using the merge sort. Finally, the two sub-arrays are merged into one single sorted array.
- The bubble sort, also known as exchange sort, is a very simple sorting method. It works by repeatedly moving the largest element to the highest position of the array.
- Quicksort is an algorithm that selects a pivot element and rearranges the values in such a way that all the elements less than the pivot element appear before it and the elements greater than the pivot appears after it.
- External sorting is a sorting technique that is used when the amount of data is massive.

6.8   EXERCISES

6.8.1 Theory Questions

Q1.   Define sorting. Write about the importance of sorting.

Q2.   What are the different types of sorting techniques? Discuss each of them in detail.

Q3.   Discuss the limitations and advantages of the insertion sort.

Q4.   Explain how a bubble sort works with a suitable example. Why is this method called a bubble sort?

Q5.   Define searching. Which search technique should be used to search for an element in an array?

Q6.   How is a linear search used to find an element? Explain how an insertion sort works with a suitable example.

Q7. Explain different types of searching techniques. Give a suitable example to illustrate a binary search.

Q8. Why is Quicksort known as "quick"?

Q9. Explain the concept of external sorting.

Q10. Differentiate between the binary search and interpolation search. Give a suitable example.

6.8.2 Programming Questions

Q1. Write a Python program to implement the bubble sort technique.

Q2. Write an algorithm to implement the interpolation search technique.

Q3. Write an algorithm to perform a merge sort. Show various stages in merge sorting over the following data: 11, 2, 9, 13, 57, 25, 17, 1, 90, 3.

Q4. Write a Python program to implement an insertion sort.

Q5. Write a program to search for an element using the binary search technique.

Q6. Write a Python program to perform a comparison sort.

Q7. Write an algorithm to perform a partition exchange sort technique. Show the various stages of the following data: 24, 52, 98, 12, 45, 6, 59, and 90.

Q8. Write an algorithm/program to implement a linear search technique.

6.8.3 Multiple Choice Questions

Q1. A binary search algorithm can be applied to a _____.

  a. Sorted array

  b. Sorted linked list

  c. Unsorted linked list

  d. Binary trees

Q2. The time complexity of a bubble sort algorithm is

  a. $O(\log n)$

  b. $O(n)$

  c. $O(n.\log n)$

  d. $O(n^2)$

Q3. Which sorting algorithm is known as a partition exchange sort?

    a. Selection Sort

    b. Merge Sort

    c. Quicksort

    d. Bubble Sort

Q4. Which case would exist when the element to be searched for using a linear search is equal to the first element of the array?

    a. Best Case

    b. Worst Case

    c. Average Case

    d. None of these

Q5. Quicksort is faster than _____.

    a. Bubble Sort

    b. Selection Sort

    c. Insertion Sort

    d. All of the above

Q6. When the amount of data is massive, which type of sorting is preferred?

    a. Internal Sorting

    b. External Sorting

    c. Both of these

    d. None of these

Q7. Which of the searching techniques is best when the value to be searched for is present in the middle?

    a. Linear Search

    b. Interpolation Search

    c. Binary Search

    d. All of these

Q8. The complexity of a binary search algorithm is _____.

    a. $O(n^2)$

    b. $O(\log n)$

    c. $O(n)$

    d. $O(n \log n)$

Q9. The selection sort has a linear running time complexity.

    a. True

    b. False

    c. Not possible to comment

7

STACKS

7.1 INTRODUCTION

A stack is an important data structure that is widely used in many computer applications. A stack can be visualized with familiar examples from our everyday lives. A very simple illustration of a stack is a pile of books, where one book is placed on top of another as shown in Figure 7.1. When we want to remove a book, we remove the topmost book first. Hence, we can add or remove an element (i.e., a book) only at or from one position, that is, the topmost position. In a stack, the element in the last position is served first. Thus, a stack can be described as a LIFO (Last In, First Out) data structure; that is, the element that is inserted last will be the first one to be taken out.

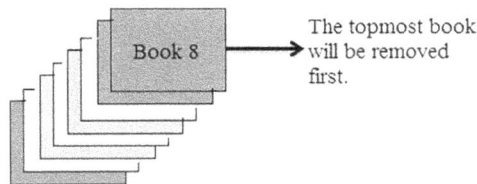

Book 8

The topmost book will be removed first.

FIGURE 7.1 A stack of books

7.2 DEFINITION OF A STACK

A *stack* is a linear collection of data elements in that the element inserted last will be the element taken out first (i.e., a stack is a LIFO data structure). The stack is an abstract data structure, somewhat similar to queues. Unlike queues, a stack is open only on one end. A stack is a linear data structure in that the insertion and deletion of elements are done only from the end called TOP. One end is always closed, and the other end is used to insert and remove data.

Stacks can be implemented by using arrays or linked lists. We discuss the implementation of stacks using arrays and linked lists in this section.

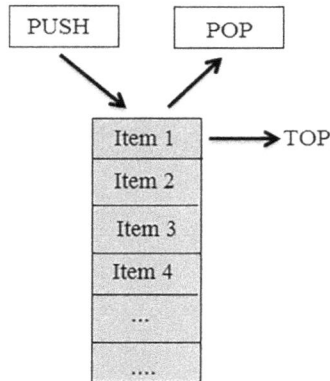

FIGURE 7.2 Representation of a stack

Practical Application:

1. A real-life example of a stack is a pile of dishes, where one dish is placed on top of another. Now, when we want to remove a dish, we remove the topmost dish first.

2. Another real-life example of a stack is a pile of disks, where one disk is placed on top of another. Now, when we want to remove a disk, we remove the topmost disk first.

7.3 OVERFLOW AND UNDERFLOW IN STACKS

Let us discuss both overflow and underflow in stacks in detail:

1. **Overflow in stacks** – The overflow condition occurs when we try to insert elements in a stack, but the stack is already full. If an attempt is made to insert a value in a stack that is already full, an overflow message is printed. It can be checked by the following formula:

   If $TOP = MAX - 1$, where MAX is the size of the stack.

2. **Underflow in stacks** – The underflow condition occurs when we try to remove elements from a stack, but the stack is already empty. If an attempt is made to delete a value from a stack that is already empty, an underflow message is printed. It can be checked by the following formula:

   If $TOP = NULL$, where MAX is the size of the stack.

Frequently Asked Questions

Q. Define a stack and list the operations performed on stacks.

Answer:

A stack is a linear data structure in that the insertion and deletion of an element are done only from the end called TOP. It is LIFO in nature (i.e., Last In, First Out). Different operations that can be performed on stacks are

a. *Push operation*

b. *Pop Operation*

c. *Peek Operation*

7.4 OPERATIONS ON STACKS

The three basic operations that can be performed on stacks are

1. PUSH

The *push* operation is the process of adding new elements in the stack. However, before inserting any new element in the stack, we must always check for the overflow condition, which occurs when we try to insert an element in a stack that is already full. An overflow condition can be checked as follows: if TOP = MAX – 1, where MAX is the size of the stack. Hence, if the overflow condition is true, then an overflow message is displayed on the screen; otherwise, the element is inserted into the stack.

For Example – Let us take a stack that has five elements in it. Suppose we want to insert another element, 10, in it; then TOP will be incremented by 1. Thus, the new element is inserted at the position pointed to by TOP. Now, let us see how a push operation occurs in the stack in Figure 7.3.

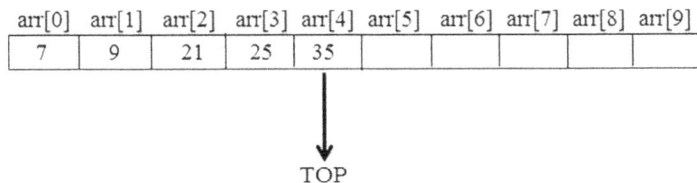

arr[0]	arr[1]	arr[2]	arr[3]	arr[4]	arr[5]	arr[6]	arr[7]	arr[8]	arr[9]
7	9	21	25	35					

TOP

After inserting 10 in it, the new stack will be

arr[0]   arr[1]   arr[2]   arr[3]   arr[4]   arr[5]   arr[6]   arr[7]   arr[8]   arr[9]

7	9	21	25	35	10				

TOP

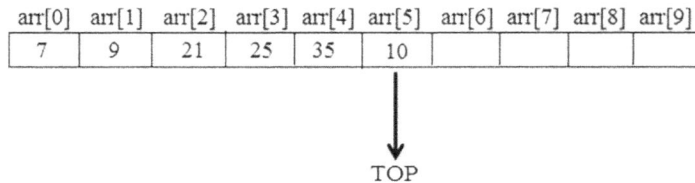

FIGURE 7.3 Stack after inserting a new element

Algorithm for a Push Operation in a Stack

```
Step 1: START
Step 2: IF TOP = MAX - 1
          Print OVERFLOW ERROR
Go to Step 5
[End of If]
Step 3: Set TOP = TOP + 1
Step 4: Set STACK[TOP] = ITEM
Step 5: EXIT
```

In the previous algorithm, we check for the overflow condition. In Step 3, TOP is incremented so that it points to the next location. Finally, the new element is inserted in the stack at the position pointed to by TOP.

2. POP

The *pop* operation is the process of removing elements from a stack. However, before deleting an element from a stack, we must always check for the underflow condition, that occurs when we try to delete an element from a stack that is already empty. An underflow condition can be checked as follows: if TOP = NULL. Hence, if the underflow condition is true, then an underflow message is displayed on the screen; otherwise, the element is deleted from the stack.

For Example – Let us take a stack that has five elements in it. Suppose we want to delete an element, 35, from the stack; then TOP will be decremented by 1. Thus, the element is deleted from the position pointed to by TOP. Now, let us see how the pop operation occurs in the stack in the following figure:

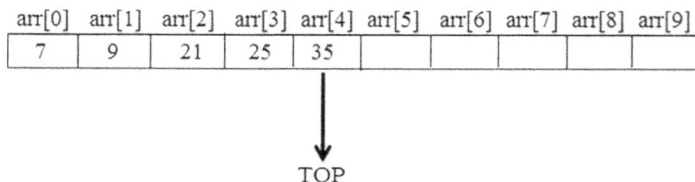

arr[0]   arr[1]   arr[2]   arr[3]   arr[4]   arr[5]   arr[6]   arr[7]   arr[8]   arr[9]

7	9	21	25	35					

TOP

After deleting 35 from it, the new stack will be

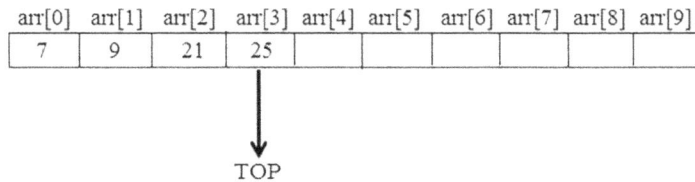

arr[0]	arr[1]	arr[2]	arr[3]	arr[4]	arr[5]	arr[6]	arr[7]	arr[8]	arr[9]
7	9	21	25						

TOP

FIGURE 7.4 A stack after deleting an element

Algorithm for the Pop Operation in a Stack

```
Step 1: START
Step 2: IF TOP = NULL
            Print UNDERFLOW ERROR
            Go to Step 5
[End of If]
Step 3: Set ITEM = STACK[TOP]
Step 4: Set TOP = TOP - 1
Step 5: EXIT
```

In the previous algorithm, first, we check for the underflow condition, that is, whether the stack is empty. If the stack is empty, then no deletion takes place; otherwise, TOP is decremented to the previous position in the stack. Finally, the element is deleted from the stack.

3. PEEK

Peek is an operation that returns the value of the topmost element of the stack. It does so without deleting the topmost element of the array. However, the peek operation first checks for the underflow condition. An underflow condition can be checked as follows: if TOP = NULL. Hence, if the underflow condition is true, then an underflow message is displayed on the screen; otherwise, the value of the element is returned.

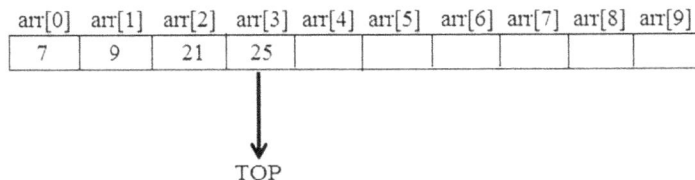

arr[0]	arr[1]	arr[2]	arr[3]	arr[4]	arr[5]	arr[6]	arr[7]	arr[8]	arr[9]
7	9	21	25						

TOP

FIGURE 7.5 Stack returning the topmost value

Algorithm for the Pop Operation in a Stack

```
Step 1: START
Step 2: IF TOP = NULL
            Print UNDERFLOW ERROR
            Go to Step 4
[End of If]
Step 3: Return STACK[TOP]
Step 4: EXIT
```

Here is a menu-driven program for stacks performing all the operations.

```python
# Python program to
# demonstrate stack implementation
# using a list

# Initializing a stack
stack = []

#function to display stack
def display():
    print(stack)
#function to delete element from stack
def pop():
    temp=stack[-1]
    stack.pop()
    print(temp,"is deleted")
#function to insert element in stack
def push():
    data=input("enter data to be insert-")
    stack.append(data)
    print("success")
#menu for stack operations
while(1):
    print(" menu ")
    print("1-push")
    print("2-pop")
    print("3-display")
    print("4-exit")
    choice=input("enter choice-")
    if choice=="1":
        push()
 elif choice=="2":
        pop()
 elif choice=="3":
        display()
 elif choice=="4":
        exit(0)
```

The output of the program is as follows:

```
    menu
1-push
2-pop
3-display
4-exit
enter choice-1
enter data to be insert-3
success
    menu
1-push
2-pop
3-display
4-exit
enter choice-1
enter data to be insert-6
success
```

```
    menu
1-push
2-pop
3-display
4-exit
enter choice-1
enter data to be insert-7
success
    menu
1-push
2-pop
3-display
4-exit
enter choice-3
['3', '6', '7']
```

```
    menu
1-push
2-pop
3-display
4-exit
enter choice-2
7 is deleted
    menu
1-push
2-pop
3-display
4-exit
enter choice-3
['3', '6']
```

Explanation:- The above menu-driven program has the push, pop, and display functions for a linear stack.

- The push function adds an element to the stack.
- The pop function removes an element from the stack.
- The display function prints every node of the stack.

7.5   IMPLEMENTATION OF A STACK

Stacks can be represented by two data structures:

1. using arrays

2. using a linked list

7.5.1 Implementation of Stacks Using Arrays

Stacks can be easily implemented using arrays. Initially, the TOP of the stack points at the first position or location of the array. As we insert new elements into the stack, the TOP keeps on incrementing, always pointing to the position where the next element will be inserted. The representation of a stack using an array is shown as follows:

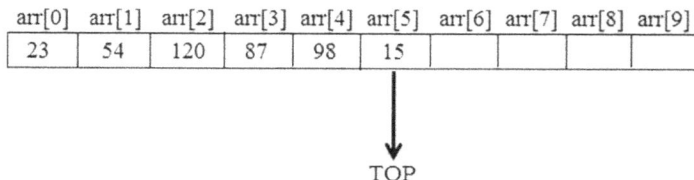

arr[0]	arr[1]	arr[2]	arr[3]	arr[4]	arr[5]	arr[6]	arr[7]	arr[8]	arr[9]
23	54	120	87	98	15				

TOP

FIGURE 7.6 Array representation of a stack

7.5.2 Implementation of Stacks Using Linked Lists

We studied how a stack is implemented using an array. Now let us discuss the same using linked lists. We already know that in linked lists, dynamic memory allocation takes place; that is, the memory is allocated at runtime. But in the case of arrays, memory is allocated at the start of the program. If we are aware of the maximum size of the stack in advance, then the implementation of stacks using arrays is efficient. But if the size is not known in advance, then we use the concept of a linked list in that dynamic memory allocation takes place. A linked list has two parts: the first part contains the information of the node and the second part stores the address of the next element in the linked list. Similarly, we can also implement a linked stack. Now, the START in the linked list becomes the TOP in a linked stack. All the insertion and deletion operations are done at the node pointed to by TOP only.

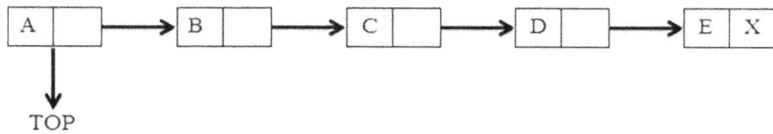

FIGURE 7.7 Linked representation of a stack

7.5.2.1 Push Operation in Linked Stacks

The push operation is the process of adding new elements in the already exist-ing stack. The new elements in the stack will always be inserted at the top-most position of the stack. Initially, we will check whether TOP = NULL. If the condition is true, then the stack is empty; otherwise, the new memory is allocated for the new node. We can understand it further with the help of an algorithm.

Algorithm for Inserting a New Element in a Linked Stack

```
Step 1: START
Step 2: Set NEW NODE. INFO = VAL
        IF TOP = NULL
Set NEW NODE. NEXT = NULL
          Set TOP = NEW NODE
ELSE
Set NEW NODE. NEXT = TOP
          Set TOP = NEW NODE
[End of If]
Step 3: EXIT
```

For Example – Consider a linked stack with four elements; a new element is to be inserted in the stack.

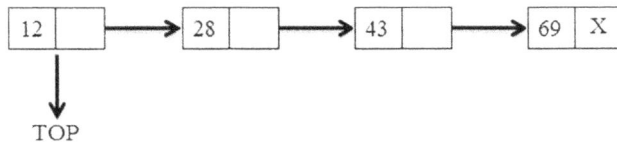

FIGURE 7.8 Linked stack before insertion

After inserting the new element in the stack, the updated stack becomes as shown in the following figure.

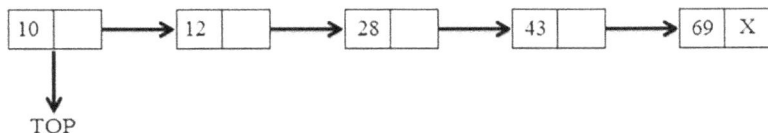

FIGURE 7.9 Linked stack after inserting a new node

7.5.2.2 Pop Operation in Linked Stacks

The pop operation is the process of removing elements from an already existing stack. The elements from the stack are always be deleted from the node pointed to by TOP. Initially, we check whether TOP = NULL. If the condition is true, then the stack is empty, which means we cannot delete any elements from it. Therefore, in that case, an underflow error message is displayed on the screen. We can understand it further with the help of an algorithm.

Algorithm for Deleting an Element from a Linked Stack

```
Step 1: START
Step 2: IF TOP = NULL
          Print UNDERFLOW ERROR
[End of If]
Step 3: Set TEMP = TOP
Step 4: Set TOP = TOP. NEXT
Step 5: FREE TEMP
Step 6: EXIT
```

For Example – Consider a linked stack with five elements; an element is to be deleted from the stack.

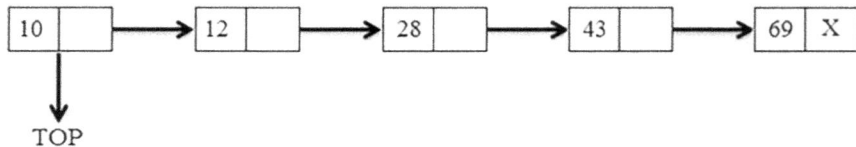

FIGURE 7.10 Linked stack before deletion

After deleting an element from the stack, the updated stack becomes as shown in Figure 7.11.

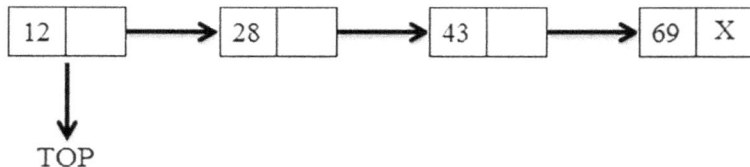

FIGURE 7.11 Linked stack after deleting the topmost node/element

Here is a program implementing a linked stack performing push and pop operations.

```
class Node:

    # Class to create nodes of linked list
    # constructor initializes node automatically
    def __init__(self,data):
        self.data = data
        self.next = None

class Stack:

    # head is default NULL
    def __init__(self):
        self.head = None

    # Checks if stack is empty
    def isempty(self):
        if self.head == None:
            return True
        else:
            return False

    # Method to add data to the stack
    # adds to the start of the stack
    def push(self,data):

        if self.head == None:
            self.head=Node(data)

        else:
            newnode = Node(data)
            newnode.next = self.head
            self.head = newnode

    # Remove element that is the current head (start of the stack)
    def pop(self):

        if self.isempty():
            return None

        else:
            # Removes the head node and makes
            #the preceding one the new head
            poppednode = self.head
            self.head = self.head.next
            poppednode.next = None
            return poppednode.data
```

```
        # Returns the head node data
        def peek(self):

            if self.isempty():
                return None

            else:
                return self.head.data

        # Prints out the stack
        def display(self):

            iternode = self.head
            if self.isempty():
                print("Stack Underflow")

            else:

                while(iternode != None):

                    print(iternode.data,"->",end = " ")
                    iternode = iternode.next
                return
```

The output of the program is as follows:

```
>>> stack=Stack()
>>> stack.push(5)
>>> stack.push(7)
>>> stack.push(9)
>>> stack.push(8)
>>> stack.display()
8 -> 9 -> 7 -> 5 ->
>>> stack.pop()
8
>>> stack.display()
9 -> 7 -> 5 ->
```

Explanation: The above program has the push, pop, peek, and display functions for a linked stack.

- The push function adds an element to the queue.
- The pop function removes elements from the queue.
- The peek function returns the topmost element of the stack.
- The display function prints every node of the queue.

7.6 APPLICATIONS OF STACKS

In this section, we discuss various applications of stacks. The topics covered in this section are as follows:

- Polish and Reverse Polish Notations
- Conversion from the Infix Expression to the Postfix Expression
- Conversion from the Infix Expression to the Prefix Expression
- Evaluation of the Postfix Expression
- Evaluation of the Prefix Expression
- Parentheses Balancing

7.6.1 Polish and Reverse Polish Notations

a. Polish Notations

Polish notation refers to a notation where the operator is placed before the operands. Polish notation was named after the Polish mathematician Jan Lukasiewicz. We can also say that transforming an expression into this form is called Polish notation. An algebraic expression can be represented in three forms. All these forms refer to the relative position of the operators in regards to the operands.

1. **Prefix Form** – In an expression, if the operator is placed before the operands, that is, +XY, then it is said to be in prefix form.

2. **Infix Form** – In an expression, if the operator is placed in the middle of the operands, that is, X + Y, then it is said to be in infix form.

3. **Postfix Form** – In an expression, if the operator is placed after the operands, that is, XY+, and then it is said to be in postfix form.

b. Reverse Polish Notation

This notation frequently refers to the postfix notation or suffix notation. It refers to the notation in that the operator is placed after its two operands, that is, XY + AF BC*.

c. The Need for Polish and Reverse Polish Notation

It is comparatively easy for a computer system to evaluate an expression in Polish notation as the system need not check for priority-wise execution of various operators (like the BODMAS rule), as all the operators in prefix or postfix expressions will automatically occur in their order of priority.

7.6.2 Conversion from the Infix Expression to the Postfix Expression

In any expression, we observe that there are two types of parts/components put together. They are operands and operators. The operators indicate the operation to be carried out, and the operands are those things on which the operators operate. Operators have their priority of execution. For the simplicity of the algorithm, we use only the addition (+), subtraction(−), modulus (%), multiplication (∗), and division (/) operators. The precedence of these operators is given as follows:

∗, ^, /, % (Higher priority)   +, − (Lower priority)

The order of evaluation of these operators can be changed by using parentheses. For example, an expression X ∗ Y + Z can be solved, as first X ∗ Y will be done and then the result is added to Z. But if the same expression is written with parentheses as X ∗ (Y + Z), now Y + Z will be evaluated first, and then the result is multiplied by X.

We can convert an infix expression to a postfix expression using a stack. First, we start to scan the expression from the left side to the right side. In an expression, there may be some operators, operands, and parentheses. Hence, we have to keep in mind some of the basic rules, that are as follows:

- Each time we encounter an operand, it is added directly to the postfix expression.

- Each time we get an operator, we should always check the top of the stack to check the priority of the operators.

- If the operator at the top of the stack has higher precedence or the same precedence as that of the current operator, then, in that case, it is repeatedly popped out from the stack and added to the postfix expression. Otherwise, it is pushed into the stack.

- Each time an opening parenthesis is encountered, it is directly pushed into the stack, and similarly, if a closing parenthesis is encountered, we repeatedly pop it out from the stack and add the operators in the postfix expression. The opening parenthesis is deleted from the stack.

Now, let us understand it with the help of an algorithm in that the first step is to push a left parenthesis in the stack and also add a closing parenthesis at the end of the infix expression. The algorithm is repeated until the stack becomes empty.

Algorithm to Convert an Infix Expression into a Postfix Expression

```
Step 1: START
Step 2: Add ")" (open parenthesis) to the end of the infix
expression.
Step 3: Push ")" on the stack.
Step 4: Repeat the steps until each character in the infix
expression is scanned.
     a) IF "('' is found, push it onto the stack.
     b) If an operand is encountered, add it to the postfix
        expression.
     c) IF ")" (close parenthesis) is found, then follow
        these steps –
        -  Continually pop from the stack and add it to the
           postfix expression until an "("is encountered.
        -  Eliminate the "(''.
     d) If an operator is found, then follow these steps –
        -  Continually pop from the stack and add it to the
           postfix expression that has the same or high
           precedence than the current operator.
        -  Push the current operator to the stack.
Step 5: Continually pop from the stack to the postfix
expression until the stack becomes empty.
Step 6: EXIT
```

For Example – Convert the following infix expression into a postfix expression.

a. $(A + B) * C / D$ **b.** $[((A + B) * (C - D)) + (F - G)]$

Solution:

a.

Character	Stack	Expression
(	(	
A	(	A
+	(+	A
B	(+	AB
)		AB+
*	*	AB+
C	*	AB+C
/	/	AB+C*
D		AB+C*D/
		Answer = AB+C*D/

b.

Character	Stack	Expression
[	[	
(	[(	
(	[((	
A	[((	A
+	[((+	A
B	[((+	AB
)	[(	AB+
*	[(*	AB+
(	[(*(	AB+
C	[(*(	AB+C
–	[(*(–	AB+C
D	[(*(–	AB+CD
)	[(*	AB+CD*
)	[	AB+CD–*
+	[+	AB+CD–*
(	[+(	AB+CD–*
F	[+(	AB+CD–*F
–	[+(–	AB+CD–*F
G	[+(–	AB+CD–*FG
)	[+	AB+CD–*FG–
]		AB+CD–*FG–+
		Answer = AB+CD–*FG–+

Here is a program to convert an infix expression to a postfix expression.

```python
# Python program to convert an infix expression to postfix

# Class to convert the expression
class Conversion:

    # Constructor to initialize the class variables
    def __init__(self, capacity):
        self.top = -1
        self.capacity = capacity
        # This array is used for a stack
        self.array = []
        # Precedence setting
        self.output = []
        self.precedence = {'+':1, '-':1, '*':2, '/':2, '^':3}

    # check if the stack is empty
    def isEmpty(self):
        return True if self.top == -1 else False

    # Return the value of the top of the stack
    def peek(self):
        return self.array[-1]

    # Pop the element from the stack
    def pop(self):
        if not self.isEmpty():
            self.top -= 1
            return self.array.pop()
        else:
            return "$"

    # Push the element to the stack
    def push(self, op):
        self.top += 1
        self.array.append(op)

    # A utility function to check is the given character
    # is operand
    def isOperand(self, ch):
        return ch.isalpha()
```

```python
        # Check if the precedence of operator is strictly
        # less than top of stack or not
        def notGreater(self, i):
            try:
                a = self.precedence[i]
                b = self.precedence[self.peek()]
                return True if a <= b else False
            except KeyError:
                return False

        # The main function that converts given infix expression
        # to postfix expression
        def infixtopostfix(self, exp):

            # Iterate over the expression for conversion
            for i in exp:
                # If the character is an operand,
                # add it to output
                if self.isOperand(i):
                    self.output.append(i)

                # If the character is an '(', push it to stack
                elif i == '(':
                    self.push(i)

                # If the scanned character is an ')', pop and
                # output from the stack until and '(' is found
                elif i == ')':
                    while( (not self.isEmpty()) and self.peek() != '('):
                        a = self.pop()
                        self.output.append(a)
                    if (not self.isEmpty() and self.peek() != '('):
                        return -1
                    else:
                        self.pop()

                # An operator is encountered
                else:
                    while(not self.isEmpty() and self.notGreater(i)):
                        self.output.append(self.pop())
                    self.push(i)

            # pop all the operator from the stack
            while not self.isEmpty():
                self.output.append(self.pop())

            print ("".join(self.output))
```

The output of the program is as follows:

```
>>> exp="a-b/(c^d+e)^(f-g*h)+i"
>>> obj=Conversion(len(exp))
>>> obj.infixtopostfix(exp)
abcd^e+fgh*-^/-i+
>>> |
```

Frequently Asked Questions

Q. Convert the following infix expression into a postfix expression.

(A + B) ^ C − (D ∗ E) / F

Answer:

Character	Stack	Expression
(	(	
A	(	A
+	(+	A
B	(+	AB
)		AB+
^	^	AB+
C	^	AB+C
−	−	AB+C^
(	−(	AB+C^
D	−(	AB+C^D
∗	−(∗	AB+C^D
E	−(∗	AB+C^DE
)	−	AB+C^DE∗
/	−/	AB+C^DE∗
F	−/	AB+C^DE∗
		Answer = AB+C^DE∗F/−

7.6.3 Conversion from an Infix Expression to a Prefix Expression

We can convert an infix expression to its equivalent prefix expression with the help of the following algorithm.

Algorithm to Convert an Infix Expression into a Prefix Expression

```
Step 1: START
Step 2: Reverse the infix expression. Also, interchange the
left and right parenthesis on reversing the infix expression.
Step 3: Obtain the postfix expression of the reversed infix
expression.
Step 4: Reverse the postfix expression so obtained in Step 3.
Finally, the expression is converted into prefix expression.
Step 5: EXIT
```

For Example – Convert the following infix expression into prefix expression.

a. $(X - Y) / (A + B)$

b. $(X - Y / Z) * (A / B - C)$

Solution:

a. After reversing the given infix expression $((B + A) / Y - X)$

Find the postfix expression of $(B + A) / (Y - X)$

Character	Stack	Expression
(	(	
(	((	
B	((	B
+	((+	B
A	((+	BA
)	(	BA+
/	(/	BA+
Y	(/	BA+Y
–	(–	BA+Y/
X	(–	BA+Y/X
)		BA+Y/X–
		BA+Y/X–

Now, reverse the postfix expression so obtained, that is, X/Y+AB.

Hence, the prefix expression is **–X/Y+AB.**

b. After reversing the given infix expression (C – B / A) * (Z / Y – X), find the postfix expression of (C – B / A) * (Z / Y – X).

Character	Stack	Expression
(	(	
C	(	C
–	(–	C
B	(–	CB
/	(–/	CB
A	(–/	CBA
)		CBA/–
*	*	CBA/–
(	*(	CBA/–
Z	*(	CBA/–Z
/	*(/	CBA/–Z
Y	*(/	CBA/–ZY
–	*(–	CBA/–ZY/
X	*(–	CBA/–ZY/X
)	*	CBA/–ZY/X–
		CBA/–ZY/X–*

Now, reverse the postfix expression so obtained, that is, *–X/ZY–/ABC.

Hence, the prefix expression is *–**X/ZY–/ABC.**

Here is a program to convert an infix expression to a prefix expression.

```python
class infix_to_prefix:
    precedence={'^':5,'*':4,'/':4,'+':3,'-':3,'(':2,')':1}
    def __init__(self):
        self.items=[]
        self.size=-1
    def push(self,value):
        self.items.append(value)
        self.size+=1
    def pop(self):
        if self.isempty():
            return 0
        else:
            self.size-=1
            return self.items.pop()
    def isempty(self):
        if(self.size==-1):
            return True
        else:
            return False
    def seek(self):
        if self.isempty():
            return False
        else:
            return self.items[self.size]
    def isOperand(self,i):
        if i.isalpha() or i in '1234567890':
            return True
        else:
            return False
    def reverse(self,expr):
        rev=""
        for i in expr:
            if i is '(':
                i=')'
            elif i is ')':
                i='('
            rev=i+rev
        return rev
    def infixtoprefix (self,expr):
        prefix=""

        for i in expr:
            if(len(expr)%2==0):
                print("Incorrect infix expr")
                return False
            elif(self.isOperand(i)):
                prefix +=i
```

```
            elif(i in '+-*/^'):
                while(len(self.items)and self.precedence[i] <
self.precedence[self.seek()]):
                    prefix+=self.pop()
                self.push(i)
            elif i is '(':
                self.push(i)
            elif i is ')':
                o=self.pop()
                while o!='(':
                    prefix +=o
                    o=self.pop()

            #end of for
        while len(self.items):
            if(self.seek()=='('):
                self.pop()
            else:
                prefix+=self.pop()

        return prefix
s=infix_to_prefix()
expr=input('enter the expression ')
rev=""
rev=s.reverse(expr)
#print(rev)
result=s.infixtoprefix(rev)
if (result!=False):

    prefix=s.reverse(result)
    print("the prefix expr of :",expr,"is",prefix)
```

The output of the program is as follows:

```
enter the expression (C - B / A) * (Z / Y - X)
the prefix expr of : (C - B / A) * (Z / Y - X) is *-C/BA-/ZYX
>>> |
```

7.6.4 Evaluation of a Postfix Expression

With the help of stacks, any postfix expression can easily be evaluated. Every character in the postfix expression is scanned from left to right. The steps involved in evaluating a postfix expression are given in the algorithm.

Algorithm for Evaluating a Postfix Expression

```
Step 1: START
Step 2: IF an operand is encountered, push it onto the stack.
Step 3: IF an operator "op1" is encountered, then follow
these steps -
        a)  Pop the two topmost elements from the stack, where X
            is the topmost element and Y is the next top element
            below X
        b)  Evaluate X op1 Y
        c)  Push the result onto the stack
Step 4: Set the result equal to the topmost element of the
stack
Step 5: EXIT
```

For Example –Evaluate the following postfix expressions.

a.   2 3 4 + * 5 6 7 8 + * + +

b.   T F T F AND F FF XOR OR AND T XOR AND OR

Solution:

a.

Character	Stack	Operation
2	2	PUSH 2
3	2, 3	PUSH 3
4	2, 3, 4	PUSH 4
+	2, 7	POP 4, 3 ADD(4 + 3 = 7) PUSH 7
*	14	POP 7, 2 MUL(7 * 2 = 14) PUSH 14

(continued)

Character	Stack	Operation
5	14, 5	PUSH 5
6	14, 5, 6	PUSH 6
7	14, 5, 6, 7	PUSH 7
8	14, 5, 6, 7, 8	PUSH 8
+	14, 5, 6, 15	POP 8, 7 ADD(8 + 7 = 15) PUSH 15
*	14, 5, 90	POP 15, 6 MUL(15 * 6 = 90) PUSH 90
+	14, 95	POP 90, 5 ADD(90 + 5 = 95) PUSH 95
+	109	POP 95, 14 ADD(95 + 14 = 109) PUSH 109
	Answer = 109	

b.

Character	Stack	Operation
T	T	PUSH T
F	T, F	PUSH F
T	T, F, T	PUSH T
F	T, F, T, F	PUSH F
AND	T, F, F	POP F, T AND(F AND T = F) PUSH F

(continued)

(continued)

Character	Stack	Operation
F	T, F, F, F	PUSH F
F	T, F, F, F, F	PUSH F
F	T, F, F, F, F, F	PUSH F
XOR	T, F, F, F, T	POP F, F XOR(F XOR F = T) PUSH T
OR	T, F, F, T	POP T, F OR(T OR F = T) PUSH T
AND	T, F, F	POP T, F AND(T AND F = F) PUSH F
T	T, F, F, T	PUSH T
XOR	T, F, F	POP T, F XOR(T XOR F = F) PUSH F
AND	T, F	POP F, F AND(F AND F = F) PUSH F
OR	T	POP F, T OR(F OR T = T) PUSH T
	Answer = T	

Here is a program for evaluation of a postfix expression.

```python
# Python program to evaluate the value of a postfix expression

# Class to convert the expression
class Evaluate:
```

```python
# Constructor to initialize the class variables
def __init__(self, capacity):
    self.top = -1
    self.capacity = capacity
    # This array is used a stack
    self.array = []

# check if the stack is empty
def isEmpty(self):
    return True if self.top == -1 else False

# Return the value of the top of the stack
def peek(self):
    return self.array[-1]

# Pop the element from the stack
def pop(self):
    if not self.isEmpty():
        self.top -= 1
        return self.array.pop()
    else:
        return "$"

# Push the element to the stack
def push(self, op):
    self.top += 1
    self.array.append(op)

# The main function that converts the given infix expression
# to a postfix expression
def evaluatePostfix(self, exp):

    # Iterate over the expression for conversion
    for i in exp:

        # If the scanned character is an operand
        # (number here) push it to the stack
        if i.isdigit():
            self.push(i)

        # If the scanned character is an operator,
        # pop two elements from stack and apply it.
        else:
            val1 = self.pop()
            val2 = self.pop()
            self.push(str(eval(val2 + i + val1)))
    return int(self.pop())
```

The output of the program is as follows:

```
>>> exp="231*+9-"
>>> obj=Evaluate(len(exp))
>>> obj.evaluatePostfix(exp)
-4
>>>
```

Frequently Asked Questions

Q. Evaluate the given postfix expression.

2 3 4 * 6 / +

Answer:

Character	Stack
2	2
3	2, 3
4	2, 3, 4
*	2, 12
6	2, 12, 6
/	2, 2
+	4
	Answer = 4

7.6.5 Evaluation of a Prefix Expression

There are a variety of techniques for evaluating a prefix expression. But the simplest of all the techniques are explained in the following algorithm.

Algorithm for Evaluating a Prefix Expression

```
Step 1: START
Step 2: Accept the prefix expression.
Step 3: Repeat steps 4 to 6 until all the characters have been
scanned.
Step 4: The prefix expression is scanned from the right.
Step 5: IF an operand is encountered, push it onto the stack.
Step 6: IF an operator is encountered, then follow these steps –
        a)  Pop two elements from the operand stack.
        b)  Apply the operator on the popped operands.
        c)  Push the result onto the stack.
Step 7: EXIT
```

For Example – Evaluate the given prefix expressions.

a. + - 4 6 * 9 / 10 50

b. + * * + 2 3 4 5 + 6 7

Solution:

a.

Character	Stack	Operation
50	50	PUSH 50
10	50, 10	PUSH 10
/	5	POP 10, 50 DIV(50 / 10 = 5) PUSH 5
9	5, 9	PUSH 9
*	45	POP 9, 5 MUL(5 * 9 = 45) PUSH 45
6	45, 6	PUSH 6
4	45, 6, 4	PUSH 4
–	45, 2	POP 4, 6 SUB(6 – 4 = 2) PUSH 2

(continued)

(continued)

Character	Stack	Operation
+	47	POP 2, 45 ADD(45 + 2 = 47) PUSH 47
	Answer = 47	

b.

Character	Stack	Operation
7	7	PUSH 7
6	7, 6	PUSH 6
+	13	POP 6, 7 ADD(7 + 6 = 13) PUSH 13
5	13, 5	PUSH 5
4	13, 5, 4	PUSH 4
3	13, 5, 4, 3	PUSH 3
2	13, 5, 4, 3, 2	PUSH 2
+	13, 5, 4, 5	POP 2, 3 ADD(3 + 2 = 5) PUSH 5
*	13, 5, 20	POP 5, 4 MUL(4 * 5 = 20) PUSH 20
*	13, 100	POP 20, 5 MUL(5 * 20 = 100) PUSH 100
+	113	POP 100, 13 ADD(13 + 100 = 113) PUSH 113
	Answer = 113	

Here is a program for evaluation of a prefix expression.

```python
class evaluate_prefix:
    def __init__(self):
        self.items=[]
        self.size=-1
    def isEmpty(self):
        if(self.size==-1):
            return True
        else:
            return False
    def push(self,item):
        self.items.append(item)
        self.size+=1
    def pop(self):
        if self.isEmpty():
            return 0
        else:
            self.size-=1
            return self.items.pop()
    def seek(self):
        if self.isEmpty():
            return False
        else:
            return self.items[self.size]
    def evaluate(self,expr):
        for i in reversed(expr):
            if i in '0123456789':
                self.push(i)
            else:
                op1=self.pop()
                op2=self.pop()
                result=self.cal(op1,op2,i)
                self.push(result)
        return self.pop()
    def cal(self,op1,op2,i):
        if i is '*':
            return int(op1)*int(op2)
        elif i is '/':
            return int(op1)/int(op2)
        elif i is '+':
            return int(op1)+int(op2)
        elif i is '-':
            return int(op1)-int(op2)
        elif i is '^':
            return int(op1)^int(op2)
```

```
s=evaluate_prefix()
expr=input('enter the prefix expression')
value=s.evaluate(expr)
print('the result of prefix expression',expr,'is',value)
```

The output of the program is as follows:

```
enter the prefix expression++596^-64
the result of prefix expression ++596^-64 is 20
>>>
```

7.6.6 Parenthesis Balancing

Stacks can be used to check the validity of parentheses in any arithmetic or algebraic expression. We are already aware that in a valid expression, the parentheses or the brackets occur in pairs, that is, if a parenthesis is open, then it must also be closed in an expression. Otherwise, the expression would be invalid. For example, $(X + Y - Z$ is invalid. But $(X + Y - Z)$ is a valid expression. Hence, some key points are to be kept in mind:

- Each time a "(" is encountered, it should be pushed onto the stack.
- Each time a ")" is encountered, the stack is examined.
- If the stack is already an empty stack, then the ")" does not have a "(", and hence the expression is invalid.
- If the stack is not empty, then we will pop the stack and check whether the popped element corresponds to the ")".
- When we reach the end of the stack, the stack must be empty. Otherwise, one or more "(" does not have a corresponding ")" and, therefore, the expression will become invalid.

For Example – Check whether the following given expressions are valid.

a. $((A - B) * Y$

b. $[(A + B) - \{X + Y\} * [C - D]]$

Solution:

a.

	Symbol	Stack
1.	(	(
2.	(	(, (
3.	A	(, (
4.	–	(, (
5.	B	(, (
6.	)	(
7.	*	(
8.	Y	(
9.		(

Answer – As the stack is not empty, the expression is not a valid expression.

b.

	Symbol	Stack
1.	[	[
2.	(	[, (
3.	A	[, (
4.	+	[, (
5.	B	[, (
6.	)	[
7.	–	[
8.	{	[, {
9.	X	[, {
10.	+	[, {
11.	Y	[, {
12.	}	[
13.	*	[
14.	[	[, [
15.	C	[, [
16.	–	[, [
17.	D	[, [
18.	]	[
19.	]	

Answer – As the stack is empty, the given expression is a valid expression.

Here is a program to implement parentheses balancing.

```python
# Python3 code to check for
# balanced parentheses in an expression
def check(my_string):
    brackets = ['()', '{}', '[]']
    while any(x in my_string for x in brackets):
        for br in brackets:
            my_string = my_string.replace(br, '')
    return not my_string

# Driver code
string =input("enter expression-")
print(string, "-", "Balanced"
      if check(string) else "Unbalanced")
```

The output of the program is as follows:

```
enter expression-[({})]
[({})] - Balanced
>>> |
```

7.7   SUMMARY

- A stack is a linear collection of data elements in that the element inserted last will be the element taken out first (i.e., a stack is a LIFO data structure). The stack is a linear data structure, in that the insertion as well as the deletion of an element, is done only from the end called TOP.

- In computer memory, stacks can be implemented by using either arrays or linked lists.

- The overflow condition occurs when we try to insert the elements in the stack, but the stack is already full.

- The underflow condition occurs when we try to remove the elements from the stack, but the stack is already empty.

- The three basic operations that can be performed on the stacks are push, pop, and peek operations.
- The push operation is the process of adding new elements in the stack.
- A pop operation is a process of removing elements from the stack.
- A peek operation is the process of returning the value of the topmost element of the stack.
- Polish notation refers to a notation where the operator is placed before the operands.
- Infix, prefix, and postfix notations are three different but equivalent notations of writing algebraic expressions.

7.8 EXERCISES

7.8.1 Theory Questions

Q1. What is a stack? Give a real-life example.

Q2. What do you understand about stack overflow and stack underflow?

Q3. What is a linked stack, and how it is different from a linear stack?

Q4. Discuss various operations that can be performed on stacks.

Q5. Explain the terms Polish notation and reverse Polish notation.

Q6. What are the various applications of a stack? Explain in detail.

Q7. Why is a stack known as a Last-In-First-Out structure?

Q8. What are different notations to represent an algebraic expression? Which one is mostly used in computers?

Q9. Explain the concept of linked stacks and also discuss how insertion and deletion take place in it.

Q10. Draw the stack structure when the following operations are performed one after another on an empty stack.

  a. Push 1, 2, 6, 17, 100

  b. Pop three numbers

  c. Peek

  d. Push 50, 23, 198, 500

  e. Display

Q11. Convert the following infix expressions to their equivalent postfix expressions.

    a. A + B + C – D * E / F

    b. [A – C] + {D * E}

    c. [X / Y] % (A * B) + (C % D)

    d. [(A – C + D) % (B – H + G)]

    e. 18 / 9 * 3 – 4 + 10 / 2

Q12. Check the validity of the given algebraic expressions.

    a. ((([A – V – D] + B)

    b. [(X – {Y * Z})]

    c. [A + C + E)

Q13. Convert the following infix expressions to their equivalent prefix expressions.

    a. 18 / 9 * 3 – 4 + 10 / 2

    b. X * (Z / Y)

    c. [(A + B) – (C + D)] * E

Q14. Evaluate the given postfix expressions.

    a. 1 2 3 * * 4 5 6 7 + + * *

    b. 12 4 / 45 + 2 3 * +

7.8.2 Programming Questions

    Q1. Write a Python program to implement a stack using arrays.

    Q2. Write a program to convert an infix expression to a prefix expression.

    Q3. Write a program to copy the contents from one stack to another using class.

    Q4. Write a Python program to convert the expression "x + y" into "xy+" using classes.

    Q5. Write a program to evaluate a postfix expression.

    Q6. Write a program to evaluate a prefix expression.

    Q7. Write a program to convert "b – c" into "–bc" using classes.

    Q8. Write a function that performs a push operation in a linked stack.

7.8.3 Multiple Choice Questions

Q1. New elements in the stack are always inserted from the

    a. Front end

    b. Top end

    c. Rear end

    d. Both (a) and (c)

Q2. A stack is a _____ data structure.

    a. FIFO

    b. LIFO

    c. FILO

    d. LILO

Q3. The overflow condition in the stack exists when

    a. TOP = NULL

    b. TOP = MAX

    c. TOP = MAX – 1

    d. None of the above

Q4. The function that inserts the elements in a stack is called _____.

    a. Push()

    b. Peek()

    c. Pop()

    d. None of the above

Q5. Disks piled up one above the other represent a _____.

    a. Queue

    b. Stack

    c. Tree

    d. Linked List

Q6. Reverse Polish notation is the other name for a _____.

    a. Postfix expression

    b. Prefix expression

    c. Infix expression

    d. All of the above

Q7. Stacks can be represented by

    a. Linked lists only

    b. Arrays only

    c. Both a) and b)

    d. None of the above

Q8. If the numbers 10, 45, 13, 50, and 32 are pushed onto a stack, what does pop return?

    a. 10

    b. 45

    c. 50

    d. 32

Q9. The postfix representation of the expression $(2 - b) * (a + 10) / (c * 8)$ is

    a. $8\, a * c\, 10 + b\, 2 - * /$

    b. $/\, 2\, a\, c * + b\, 10 * 9 -$

    c. $2\, b - a\, 10 + * c\, 8 * /$

    d. $10\, a + * 2\, b - / c\, 8 *$

## CHAPTER 8

TREES

8.1 INTRODUCTION

In earlier chapters, we learned about various data structures such as arrays, linked lists, stacks, and queues. All these data structures are linear data structures. Although linear data structures are flexible, it is quite difficult to use them to organize data into a hierarchical representation. Hence, to overcome this problem or limitation, we create a new data structure that is called a tree. A tree is a data structure that is defined as a set of one or more nodes that allows us to associate a parent-child relationship. In trees, one node is designated as the root node or parent node, and all the remaining nodes can be partitioned into non-empty sets, each of which is a sub-tree of the root. Unlike natural trees, a tree data structure is upside down, having a root at the top and leaves at the bottom. Also, there is no parent of the root node. A root node can only have child nodes. Leaf nodes or leaves are those that have no children. When there are no nodes in the tree, then the tree is known as a *null tree* or *empty tree*. Trees are widely used in various day-to-day applications. The recursive programming of trees makes the programs optimized and easily understandable. Trees are also used to represent the structure of mathematical formulas. Figure 8.1 represents a tree, where A is the root node of the tree. X, Y, and Z are the child nodes of the root node A. They also form the sub-trees of the tree. B, C, Y, D, and E are the leaf nodes of the tree, as they have no children.

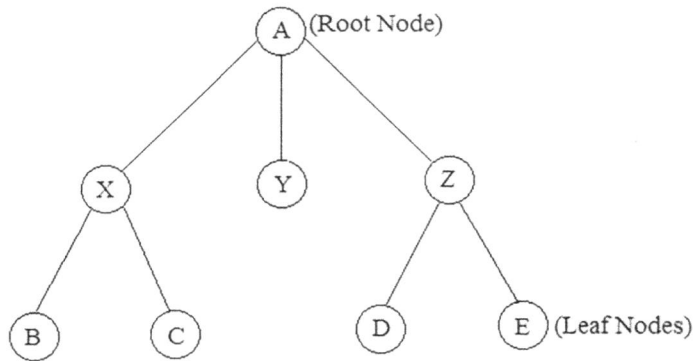

FIGURE 8.1 A tree

Practical Application:

1. The members of a family can be visualized as a tree in that the root node can be visualized as a grandfather. His two children can be visualized as the child nodes. Then the grandchildren form the left and the right sub-trees of the tree.

2. Trees are used to organize information in database systems and represent the syntactic structure of the source programs in compilers.

8.2 DEFINITIONS

- **Node** – A node is the main component of the tree data structure. It stores the actual data along with the links to the other nodes.

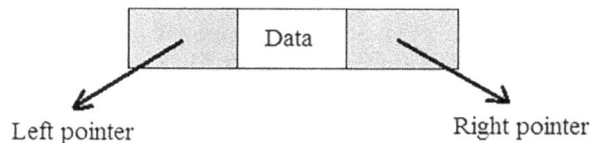

FIGURE 8.2 Structure of a node

- **Root** – The root node is the topmost node of the tree. It does not have a parent node. If the root node is empty, then the tree is empty.

- **Parent** – The parent of a node is the immediate predecessor of that node. In the following figure, X is the parent of the Y and Z nodes.

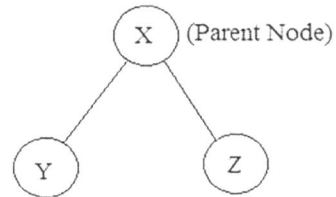

FIGURE 8.3 Parent node

- **Child** – The child nodes are the immediate successors of a node. They must have a parent node. A child node placed at the left side is called the *left child*, and similarly, a child node placed at the right side is called a *right child*. Y is the left child of X, and Z is the right child of X.

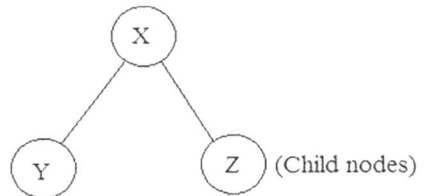

FIGURE 8.4 Child nodes

- **Leaf/Terminal nodes** – A leaf node is one that does not have any child nodes.

- **Sub-trees** – The nodes B, X, and Y form the left sub-tree of root A. Similarly, the nodes C and Z form the right sub-tree of A.

- **Path** – It is a unique sequence of consecutive edges that is required to be followed to reach the destination from a given source. The path from root node A to Y is given as A-B, B-Y.

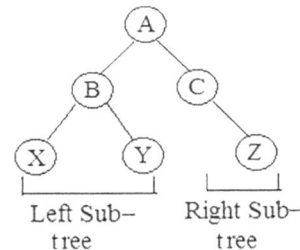

FIGURE 8.5 Sub-trees

- **Level number of a node** – Every node in the tree is assigned a level number. The root is at level 0, the children of the root node are at level 1, and so on.

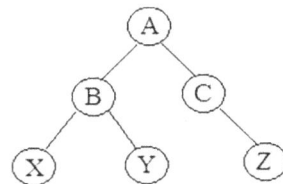

FIGURE 8.6 Path

- **Height** – The height of the tree is the maximum level of the node + 1. The height of a tree containing a single node is 1. Similarly, the height of an empty tree is 0.

- **Ancestors** – The ancestors of a node are any predecessor nodes on the path between the root and the destination. There are no ancestors for the root node. The nodes A and B are the ancestors of node X.

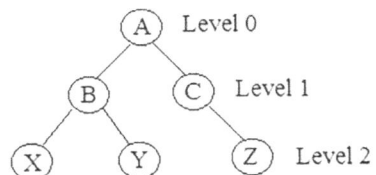

FIGURE 8.7 Node level numbers

- **Descendants** – The descendants of a node are any successor nodes on the path between the given source and the leaf node. There are no descendants of the leaf node. Here, B, X, and Y are the descendants of node A.

- **Siblings** – The child nodes of a given parent node are called siblings. X and Y are the siblings of B in Figure 8.8.

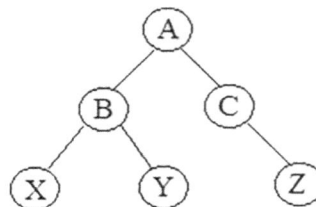

FIGURE 8.8 Siblings

- **Degree of a node** – This is equal to the number of children that a node has.

- **Out-degree of a node** – This is equal to the number of edges leaving that node.

- **In-degree of a node** – This is equal to the number of edges arriving at that node.

- **Depth** – This is given as the length of the path from the root node to the destination node.

8.3 BINARY TREE

A *binary tree* is a collection of nodes where each node contains three parts: the left child address, right child address, and the data item. The left child address stores the memory location of the top node of the left sub-tree and the right child address stores the memory location of the top node of the right sub-tree. The topmost element of the binary tree is known as a *root node*. The root stores the memory location of the root node. As the name suggests, a binary tree can have at most two children, that is, a parent can have zero, one, or at most two children. If root = NULL, then it means that the tree is empty. Figure 8.9 represents a binary tree.

In the following figure, A represents the root node of the tree. B and C are the children of root node A. Nodes B, D, E, F, and G constitute the left sub-tree. Similarly, nodes C, H, I, and J constitute the right sub-tree. Now, nodes G, E, F, I, and J are the terminal/leaf nodes of the binary tree, as they have no children. Hence, node A has two successors B and C. Node B has two successors D and G. Similarly, node D also has two successors E and F. Node G has no successor. Node C has only one successor H. Node H has two successors I and J. Since nodes E, F, G, I, and J have no successors, they are said to have empty sub-trees.

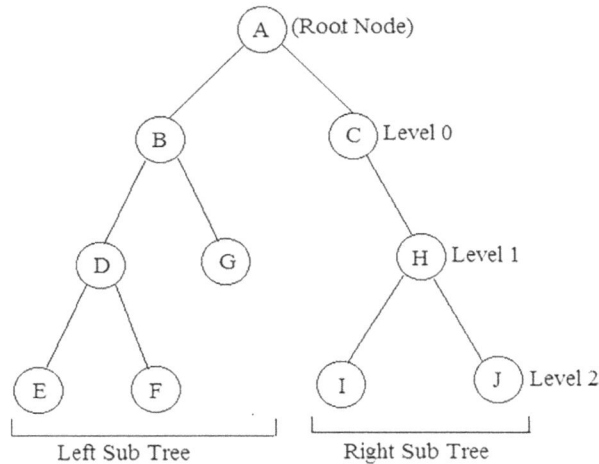

FIGURE 8.9 A binary tree

8.3.1 Types of Binary Trees

There are two types of binary trees:

1. **Complete Binary Trees** – A complete binary tree is a type of binary tree that obeys/satisfies two properties:

   a. First, every level in a complete binary tree except the last one must be completely filled.

   b. Second, all the nodes in the complete binary tree must appear left as much as possible.

   In a complete binary tree, the number of nodes at level n is 2^n nodes. The total number of nodes in a complete binary tree of depth d is equal to the sum of all nodes present at each level between 0 and d.

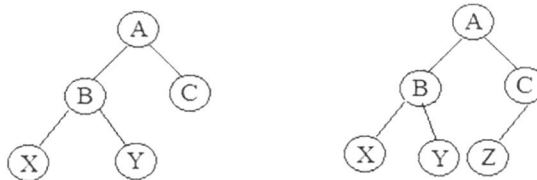

FIGURE 8.10 Complete binary trees

2. **Extended Binary Trees** – Extended binary trees are also known as 2T-trees. A binary tree is said to be an extended binary tree if, and only

if, every node in the tree has either zero children or two children. In an extended binary tree, nodes with two children are known as internal nodes. Nodes with no children are known as external nodes. In the following figure, the internal nodes are represented by I and the external nodes are represented by E.

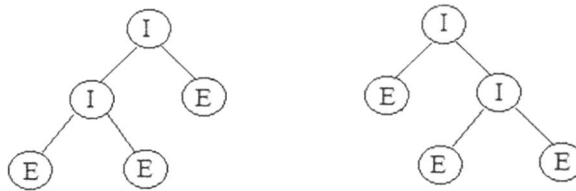

FIGURE 8.11 Extended binary trees

8.3.2 Memory Representation of Binary Trees

Binary trees can be represented in a computer's memory in either of the following ways:

1. Array/List Representation of Binary Trees
2. Linked Representation of Binary Trees

Array Representation of Binary Trees

A binary tree is represented using an array in the computer's memory. It is also known as sequential representation. Sequential representation of binary trees is done using one-dimensional (1D) arrays. This type of representation is static and hence inefficient, as the size must be known in advance and thus requires a lot of memory space. The following rules are used to decide the location of each node in the memory:

a. The root node of the tree is stored in the first location.

b. If the parent node is present at location k, then the left child is stored at location 2k, and the right child is stored at location (2k + 1).

c. The maximum size of the array is given as $(2^h - 1)$, where h is the height of the tree.

For Example – A binary tree is given as follows. Give its array representation in the memory.

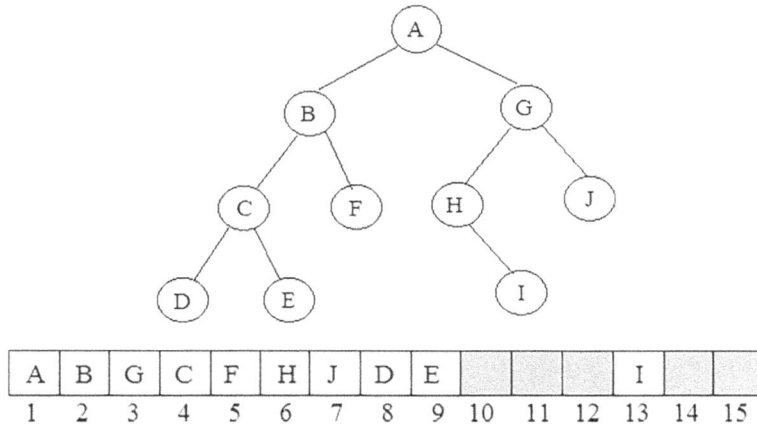

FIGURE 8.12 A binary tree and its array representation

Linked Representation of Binary Trees

A binary tree can also be represented using a linked list in a computer's memory. This type of representation is dynamic, as memory is dynamically allocated, that is, when it is needed, and thus it is efficient and avoids wastage of memory space. In linked representation, every node has three parts:

1. The first part is called the left child, which contains the address of the left sub-tree.

2. The second part is called the data part, which contains the information of the node.

3. The third part is called the right child, which contains the address of the right sub-tree.

The class of the node is declared as follows:

```
class Node:
    def __init__(self,data):
        self.data=data
        self.left=None
        self.right=None
```

The representation of a node is given in Figure 8.2. When there are no children of a node, the corresponding fields are NULL.

For Example – A binary tree is given as follows. Give its linked representation in the memory.

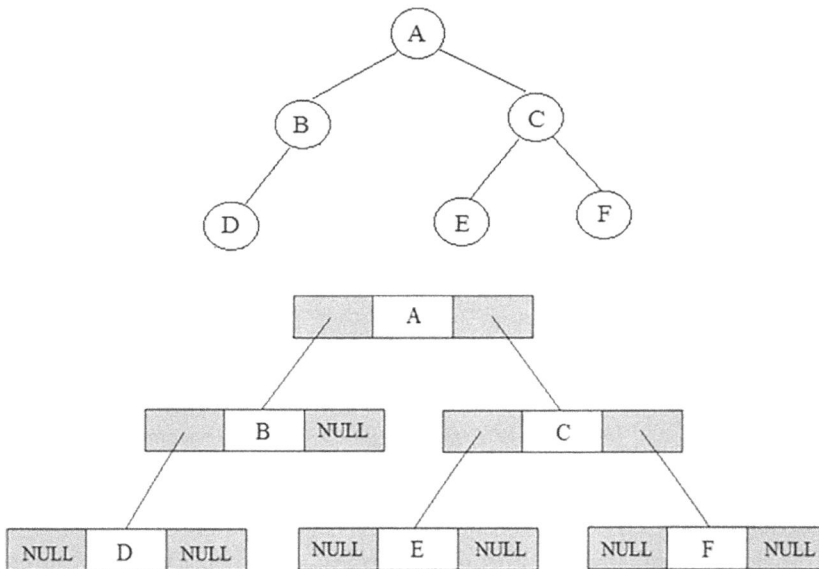

FIGURE 8.13 A binary tree and its linked representation

8.4 BINARY SEARCH TREE

A *Binary Search Tree* (BST) is a variant of a binary tree. The special property of a binary search tree is that all the nodes in the left sub-tree have a value less than that of the root node. Similarly, all the nodes in the right sub-tree have a value more than that of the root node. Hence, the binary search tree is also known as an *ordered binary tree*, because all the nodes in a binary search tree are ordered. The left and the right sub-trees are also binary search trees, and thus the same property is applicable on every sub-tree in the binary search tree. Figure 8.14 represents a binary search tree in that all the keys are ordered.

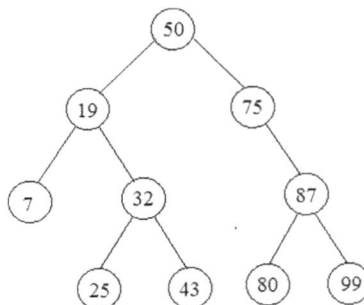

FIGURE 8.14 A binary search tree

In Figure 8.14, the root node is 50. The left sub-tree of the root node consists of the nodes 19, 7, 32, 25, and 43. We can see that all these nodes have smaller values than the root node, and hence it constitutes the left sub-tree. Similarly, the right sub-tree of the root node consists of the nodes 75, 87, 80, and 99. Here also, we can see that all these nodes have higher values than the root node and hence it constitutes the right sub-tree. Each of the sub-trees is ordered. Thus, it becomes easier to search for an element in the tree, and as a result, time is also reduced by a great margin. Binary search trees are very efficient regarding searching for an element. These trees are already sorted in nature. Thus, these trees have a low time complexity. Various operations that can be performed on binary search trees are discussed in the upcoming section.

8.4.1 Operations on Binary Search Trees

In this section, we discuss different operations that are performed on binary search trees, which include

- Searching a node/key in the binary search tree
- Inserting a node/key in the binary search tree
- Deleting a node/key from the binary search tree
- Deleting the entire binary search tree
- Finding the mirror image of the binary search tree
- Finding the smallest node in the binary search tree
- Finding the largest node in the binary search tree
- Determining the height of the binary search tree

1. **Searching a node/key in the binary search tree** – The search operation is one of the most common operations performed in the binary search tree. This operation is performed to find whether a given key exists in the tree. The search operation starts at the root node. First, it checks whether the tree is empty. If the tree is empty, then the node/key we are searching for is not present in the tree, and the algorithm terminates there by displaying the appropriate message. If the tree is not empty and the nodes are present in it, then the search function checks the node/value to be searched and compares it with the key value of the current node. If the node/key to be searched is less than the key value of the current node, then in that case, we recursively call the left child node. On the other hand, if the node/key to be searched is greater than the key value of the current node, then we recursively call the right child node. Now, let us look at the algorithm for searching for a key in a binary search tree.

Algorithm for Searching for a Node/Key in a Binary Search Tree

```
SEARCH(ROOT, VALUE)
Step 1: START
Step 2: IF(ROOT == NULL)
            Return NULL
            Print "Empty Tree"
      ELSE IF(ROOT . INFO == VALUE)
            Return ROOT
      ELSE IF(ROOT . INFO > VALUE)
            SEARCH(ROOT . LCHILD, VALUE)
      ELSE IF(ROOT . INFO < VALUE)
            SEARCH(ROOT . RCHILD, VALUE)
      ELSE
            Print "Value not found"
        [End of IF]
      [End of IF]
    [End of IF]
  [End of IF]
Step 3: END
```

In the previous algorithm, we check whether the tree is empty. If the tree is empty, then we return NULL. If the tree is not empty, then we check whether the value stored at the current node (ROOT) is equal to the node/key we want to search. If the value of the ROOT node is equal to the key value to be searched, then we return the current node of the tree, that is, the ROOT node. Otherwise, if the key value to be searched is less than the value stored at the current node, we recursively call the left sub-tree. If the key value to be searched is greater than the value stored at the current node, then we recursively call the right sub-tree. Finally, if the value is not found, then an appropriate message is printed on the screen.

For Example – We have been given a binary search tree. Now, search the node with the value 20 in the binary search tree.

Initially the binary search tree is given as

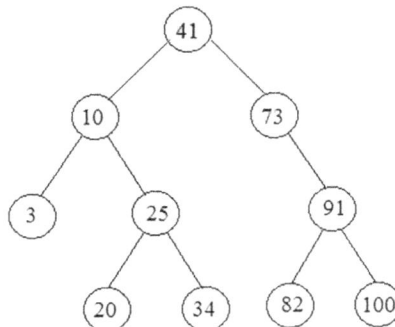

FIGURE 8.15(a)

Step 1: First, the root node, 41, is checked.

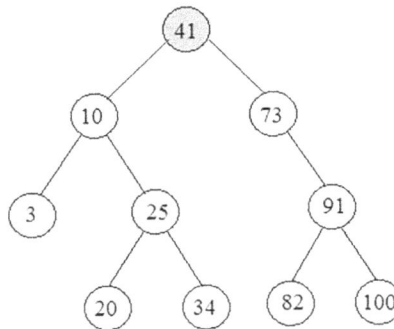

FIGURE 8.15(b)

Step 2: Second, as the value stored at the root node is not equal to the value to be searched, but we know that 20 < 41, thus we traverse the left sub-tree.

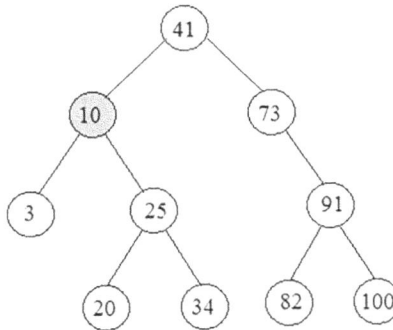

FIGURE 8.15(c)

Step 3: We know that 10 is not the value to be searched, but 20 > 10; thus, we now traverse the right sub-tree with respect to 10.

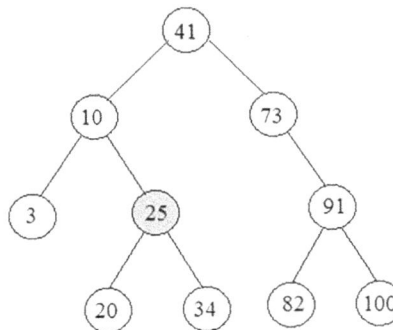

FIGURE 8.15(d)

Step 4: Again, 25 is not the value to be searched, but 20 < 25; thus, we now traverse the left sub-tree with respect to 25.

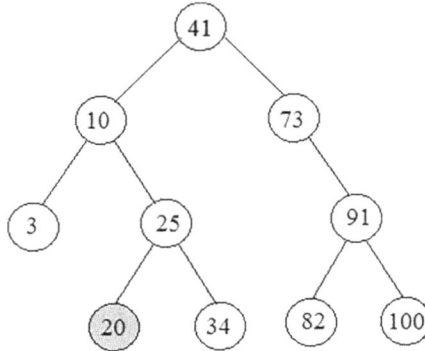

FIGURE 8.15 Searching a node with value 20 in the binary search tree

Finally, a node having value 20 is successfully searched for in the binary search tree.

2. **Inserting a node/key in the binary search tree** – The insertion operation is performed to insert a new node with the given value in the binary search tree. The new node is inserted at the correct position following the binary search tree constraint. It should not violate the property of the binary search tree. The insertion operation also starts at the root node. First, it checks whether the tree is empty. If the tree is empty, then we allocate the memory for the new node. If the tree is not empty, then we compare the key value to be inserted with the value stored in the current node. If the node/key to be inserted is less than the key value of the current node, then the new node is inserted in the left sub-tree. If the node/key to be inserted is greater than the key value of the current node, then the new node is inserted in the right sub-tree. Now, let us discuss the algorithm for inserting a node in the binary search tree.

Algorithm for Inserting a Node/Key in a Binary Search Tree

```
INSERT(ROOT, VALUE)
Step 1: START
Step 2: IF(ROOT == NULL)
            Allocate memory for ROOT node
            Set ROOT . INFO = VALUE
            Set ROOT . LCHILD = ROOT . RCHILD = NULL
        [End of IF]
```

```
Step 3: IF(ROOT . INFO > VALUE)
            INSERT(ROOT . LCHILD, VALUE)
        ELSE
            INSERT(ROOT . RCHILD, VALUE)
            [End of IF]
Step 4: END
```

In the previous algorithm, we check whether the tree is empty. If the tree is empty, then we allocate memory for the ROOT node. In Step 3, we check whether the key value to be inserted is less than the value stored at the current node; if so, we simply insert the new node in the left sub-tree. Otherwise, the new child node is inserted in the right sub-tree.

For Example – We have been given a binary search tree. Now, insert a new node with the value 7 in the binary search tree.

Initially, the binary search tree is given as

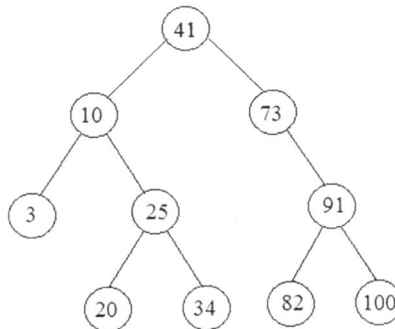

FIGURE 8.16(a)

Step 1: First, we check whether the tree is empty, so we check the root node. As the root node is not empty, we begin the insertion process.

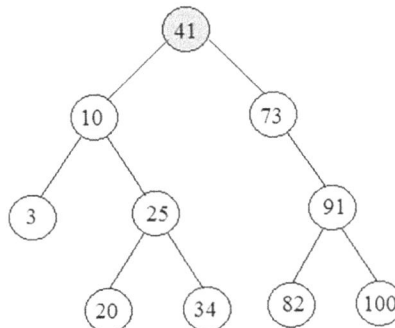

FIGURE 8.16(b)

Step 2: Second, we know that 7 < 41; thus, we traverse the left sub-tree to insert the new node.

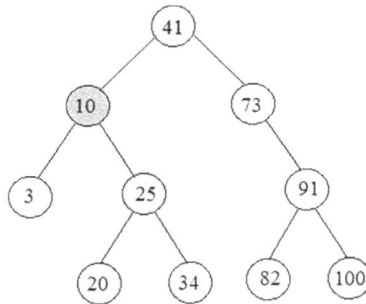

FIGURE 8.16(C)

Step 3: Third, we know that 7 < 10; thus, we again traverse the left sub-tree to insert the new node.

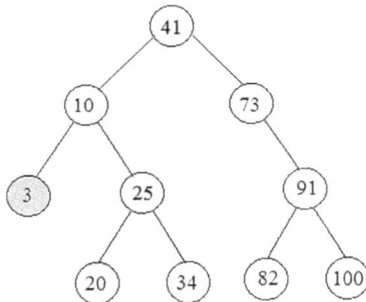

FIGURE 8.16(d)

Step 4: Now, we know that 7 > 3, thus the new node with value 7 is inserted as the right child of the parent node 3.

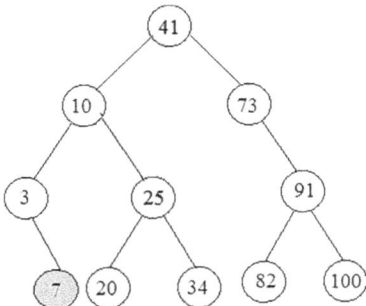

FIGURE 8.16 Inserting a new node with value 7 in the binary search tree

Finally, the new node with the value 7 is inserted as a right child in the binary search tree.

3. **Deleting a node/key from a binary search tree** – Deleting a node/key from a binary search tree is the most crucial process. We should be careful when performing the deletion operation; while deleting the nodes, we must be sure that the property of the binary search tree is not violated so that we don't lose the necessary nodes during this process. The deletion operation is divided into three cases.

Case 1: Deleting a Node Having No Children

This is the simplest case of deletion, as we can directly remove or delete a node that has no children. Look at the binary search tree given in Figure 8.17 and see how the deletion is done in this case.

For Example – We have been given a binary search tree. Now, delete a node with the value 61 from the binary search tree.

Initially the binary search tree is given as shown in Figure 8.17(a).

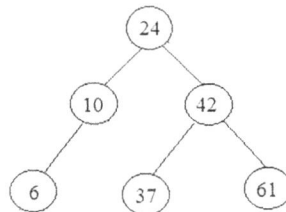

FIGURE 8.17(a)

Step 1: First, we check whether the tree is empty by checking the root node.

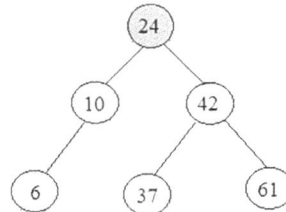

FIGURE 8.17(b)

Step 2: Second, as the root node is present, we compare the value to be deleted with the value stored at the current node. As 61 > 24, we recursively traverse the right sub-tree.

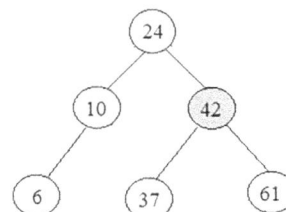

FIGURE 8.17(c)

Step 3: Again, we compare the value to be deleted with the value stored at the current node. As 61 > 42, we recursively traverse the right sub-tree.

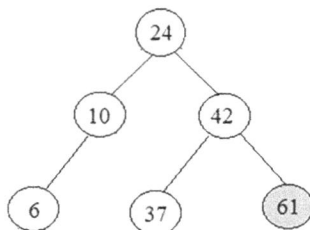

FIGURE 8.17(d)

Step 4: Finally, a node having value 61 is deleted from the binary search tree.

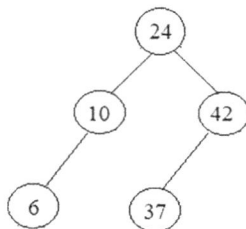

FIGURE 8.17 Deleting the node with value 61 from the binary search tree

Case 2: Deleting a Node Having One Child

In this case of deletion, the node that is to be deleted, the parent node, is simply replaced by its child node. Look at the binary search tree given in Figure 8.18 and see how the deletion is done in this case.

For Example – We have been given a binary search tree. Now, delete a node with the value 10 from the binary search tree.

Initially the binary search tree is given as

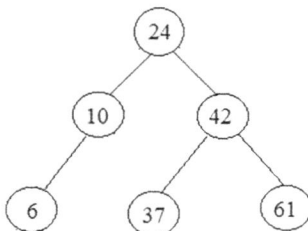

FIGURE 8.18(a)

Step 1: First, we check whether the tree is empty by checking the root node.

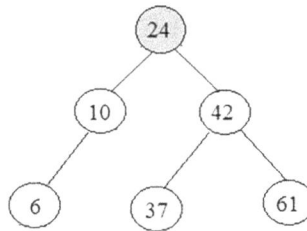

FIGURE 8.18(b)

Step 2: Second, as the root node is present, we compare the value to be deleted with the value stored at the current node. As 10 < 24, we recursively traverse the left sub-tree.

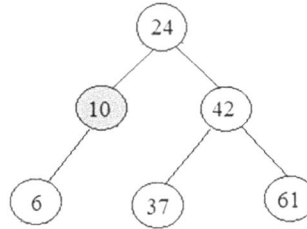

FIGURE 8.18(c)

Step 3: Now, as the node to be deleted is found and has one child, the node to be deleted is replaced by its child node, and the actual node is deleted.

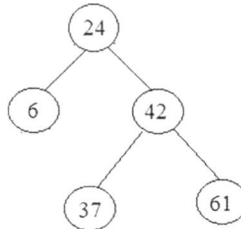

FIGURE 8.18 Deleting the node with value 10 from the binary search tree

Case 3: Deleting a Node Having Two Children

In this case, the node that is to be deleted is simply replaced by its in-order predecessor, that is, the largest value in the left sub-tree, or by its in-order successor, that is, the smallest value in the right sub-tree. The in-order predecessor or in-order successor can be deleted using any of the two cases. Look at the binary search tree shown in Figure 8.19 and see how the deletion takes place in this case.

Now, let us discuss the algorithm for deleting a node from a binary search tree.

For Example – We have been given a binary search tree. Now, delete a node with the value 42 from the binary search tree.

Initially, the binary search tree is given as

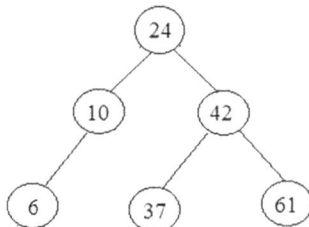

FIGURE 8.19(a)

Step 1: First, we check whether the tree is empty or not by checking the root node.

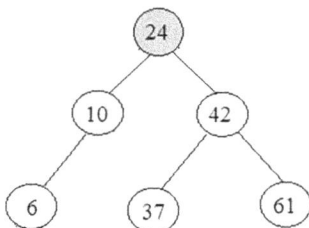

FIGURE 8.19(b)

Step 2: Second, as the root node is present, we compare the value to be deleted with the value stored at the current node. As 42 <>24, we recursively traverse the right sub-tree.

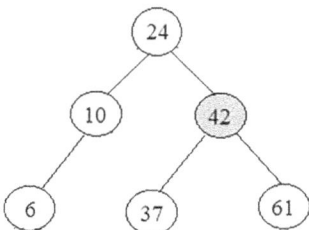

FIGURE 8.19(c)

Step 3: As the node to be deleted is found and has two children, now we find the in-order predecessor of the current node (42) and replace the current node with its in-order predecessor so that the actual node 42 is deleted.

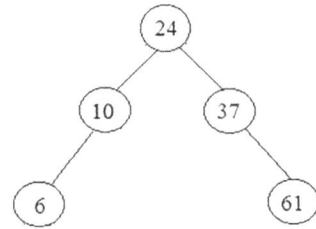

FIGURE 8.19 Deleting the node with the value 42 from the binary search tree

Algorithm for Deleting a Node/Key from a Binary Search Tree

```
DELETE_NODE(ROOT, VALUE)
Step 1: START
Step 2: IF(ROOT == NULL)
            Print "Error"
        [End of IF]
Step 3: IF(ROOT . INFO > VALUE)
            DELETE_NODE(ROOT . LCHILD, VALUE)
        ELSE IF(ROOT . INFO < VALUE)
            DELETE_NODE(ROOT . RCHILD, VALUE)
        ELSE IF(ROOT . LCHILD = NULL & ROOT . RCHILD = NULL)
        FREE(ROOT)
        ELSE
            IF(ROOT . LCHILD & ROOT . RCHILD)
TEMP = FIND_LARGEST(ROOT . LCHILD)
        OR
            TEMP = FIND_SMALLEST(ROOT . RCHILD)
            Set ROOT . INFO = TEMP . INFO
            FREE(TEMP)
        ELSE
            IF(ROOT . LCHILD != NULL)
            Set TEMP = ROOT . LCHILD
            Set ROOT . INFO = TEMP . INFO
            FREE(TEMP)
        ELSE
            Set TEMP = ROOT . RCHILD
            Set ROOT . INFO = TEMP . INFO
            FREE(TEMP)
        [End of IF]
      [End of IF]
    [End of IF]
Step 4: END
```

In the previous algorithm, we check whether the tree is empty. If the tree is empty, then the node to be deleted is not present. Otherwise, if the tree is not empty, we check whether the node/value to be deleted is less than the value stored at the current node. If the value to be deleted is less, then we recursively call the left sub-tree. If the value to be deleted is greater than the value stored at the current node, then we recursively call the right sub-tree. Now, if the node to be deleted has no children, then the node is simply freed. If the node to be deleted has two children, that is, both a left and right child, then we find the in-order predecessor by calling (TEMP = FIND_LARGEST(ROOT . LCHILD) or in-order successor by calling (TEMP = FIND_SMALLEST(ROOT . RCHILD) and replace the value stored at the current node with that of the in-order predecessor or in-order successor. Then, we simply delete the initial node of either the in-order predecessor or in-order successor. Finally, if the node to be deleted has only one child, the value stored at the current node is replaced by its child node and the child node is deleted.

4. **Deleting the entire binary search tree** – It is very easy to delete the entire binary search tree. First, we delete all the nodes present in the left sub-trees followed by the nodes present in the right sub-tree. Finally, the root node is deleted, and the entire tree is deleted.

Algorithm for Deleting an Entire Binary Search Tree

```
DELETE_BST(ROOT)

Step 1: START
Step 2: IF(ROOT != NULL)
            DELETE_BST(ROOT . LCHILD)
            DELETE_BST(ROOT . RCHILD)
            FREE(ROOT)
        [End of IF]
Step 3: END
```

5. **Finding the mirror image of a binary search tree** – This is an exciting operation to perform in a binary search tree. The mirror image of the binary search tree means interchanging the right sub-tree with the left sub-tree at each and every node of the tree.

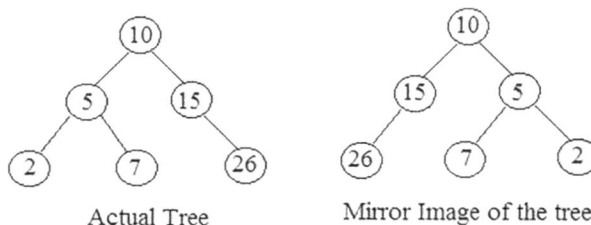

Actual Tree   Mirror Image of the tree

FIGURE 8.20 A binary search tree and its mirror image

Algorithm for Finding the Mirror Image of a Binary Search Tree

```
MIRROR_IMAGE(ROOT)

Step 1: START
Step 2: IF(ROOT != NULL)
            MIRROR_IMAGE(ROOT . LCHILD)
            MIRROR_IMAGR(ROOT . RCHILD)
            Set TEMP = ROOT . LEFT
            ROOT . LEFT = ROOT . RIGHT
            Set ROOT . RIGHT = TEMP
        [End of IF]
Step 3: END
```

6. **Finding the smallest node in the binary search tree** – We know that it is the basic property of the binary search tree that the smallest value always occurs in the extreme left of the left sub-tree. If there is no left sub-tree, then the value of the root node will be the smallest. Hence, to find the smallest value in the binary search tree, we simply find the value of the node present at the extreme left of the left sub-tree.

Algorithm for Finding the Smallest Node in a Binary Search Tree

```
SMALLEST_VALUE(ROOT)

Step 1: START
Step 2: IF(ROOT = NULL OR ROOT . LCHILD = NULL)
            Return NULL
        ELSE
            Return SMALLEST_VALUE(ROOT)
        [End of IF]
Step 3: END
```

7. **Finding the largest node in a binary search tree** – We know that it is the basic property of the binary search tree that the largest value always occurs in the extreme right of the right sub-tree. If there is no right sub-tree, the value of the root node will be the largest. Hence, to find the largest value in a binary search tree, we simply find the value of the node present at the extreme right of the right sub-tree.

Algorithm for Finding the Largest Node in a Binary Search Tree

```
LARGEST_VALUE(ROOT)

Step 1: START
Step 2: IF(ROOT = NULL OR ROOT . RCHILD = NULL)
          Return NULL
        ELSE
          Return LARGEST_VALUE(ROOT)
        [End of IF]
Step 3: END
```

8. **Determining the height of a binary search tree** – The height of a binary search tree can easily be determined. We first calculate the heights of the left sub-tree and the right sub-tree. If that height is greater, 1 is added to that height; that is, if the height of the left sub-tree is greater, then 1 is added to the height of the left sub-tree. Similarly, if the height of the right sub-tree is greater, then 1 is added to the height of the right sub-tree.

Algorithm for Determining the Height of a Binary Search Tree

```
CALCULATE_HEIGHT(ROOT)
Step 1: START
Step 2: IF ROOT = NULL
          Print "Can't find height of the tree."
        ELSE
          Set LHEIGHT = CALCULATE_HEIGHT(ROOT . LCHILD)
          Set RHEIGHT = CALCULATE_HEIGHT(ROOT . RCHILD)
        IF(LHEIGHT < RHEIGHT)
Return (RHEIGHT) + 1
      ELSE
          Return (LHEIGHT) + 1
        [End of IF]
        [End of IF]
Step 3: END
```

8.4.2 Binary Tree Traversal Methods

Traversing is the process of visiting each node in the tree exactly once in a particular order. We know that a tree is a non-linear data structure, and therefore a tree can be traversed in various ways. There are three types of traversals:

- Pre-Order Traversal
- In-Order Traversal
- Post-Order Traversal

Pre-Order Traversal

In pre-order traversal, the following operations are performed recursively at each node:

1. Visit the root node.

2. Traverse the left sub-tree.

3. Traverse the right sub-tree.

The word "pre" in pre-order determines that the root node is accessed before accessing any other node in the tree. Hence, it is also known as a DLR traversal, that is, Data Left Right. Therefore, in a DLR traversal, the root node is accessed first, followed by the left sub-tree and right sub-tree. Now, let us see an example of pre-order traversal.

For Example – Find the pre-order traversal of the given binary tree of the word EDUCATION.

The pre-order traversal of the previous binary tree is

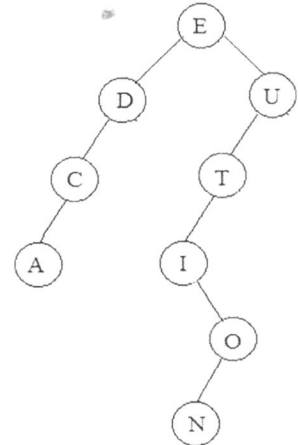

E D C A U T I O N

Now, let us look at the function for pre-order traversal.

Function for Pre-Order Traversal

```python
# A function to do preorder tree traversal
def printPreorder(root):

    if root:

        # First print the data of node
        print(root.val),

        # Then recur on left child
        printPreorder(root.left)

        # Finally recur on right child
        printPreorder(root.right)
```

In-Order Traversal

In in-order traversal, the following operations are performed recursively at each node:

1. Traverse the left sub-tree.

2. Visit the root node.

3. Traverse the right sub-tree.

The word "in" in "in-order" determines that the root node is accessed in between the left and the right sub-trees. Hence, it is also known as an LDR traversal, that is, Left Data Right. Therefore, in an LDR traversal, the left sub-tree is traversed first followed by the root node and the right sub-tree. Now, let us see an example of an in-order traversal.

For Example – Find the in-order traversal of the given binary tree of the word EDUCATION.

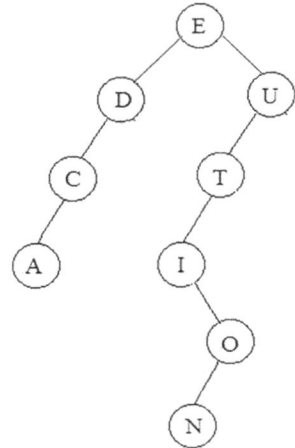

The in-order traversal of the previous binary tree is as follows:

A C D E I N O T U

Now, let us look at the function for an in-order traversal.

Function for an In-Order Traversal

```
# A function to do inorder tree traversal
def printInorder(root):

    if root:

        # First recur on left child
        printInorder(root.left)

        # then print the data of node
        print(root.val),

        # now recur on right child
        printInorder(root.right)
```

Post-Order Traversal

In a post-order traversal, the following operations are performed recursively at each node:

1. Traverse the left sub-tree

2. Traverse the right sub-tree

3. Visit the root node

The word "post" in post-order determines that the root node will be accessed last after the left and the right sub-trees. Hence, it is also known as an LRD traversal, that is, Left Right Data. Therefore, in an LRD traversal, the left sub-tree is traversed first followed by the right sub-tree and the root node. Now, let us see an example for a post-order traversal.

For Example – Find the post-order traversal of the given binary tree of the word EDUCATION.

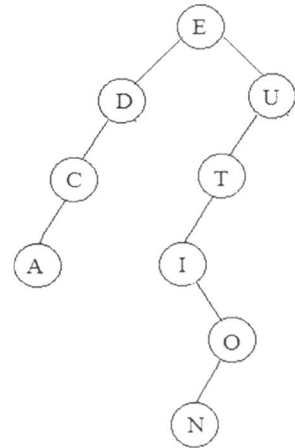

The post-order traversal of the previous binary tree is as follows:

```
A  C  D  N  O  I  T  U  E
```

Now, let us look at the function of the post-order traversal.

Function for Post-Order Traversal

```
# A function to do postorder tree traversal
def printPostorder(root):

    if root:

        # First recur on left child
        printPostorder(root.left)

        # the recur on right child
        printPostorder(root.right)

        # now print the data of node
        print(root.val)
```

Here is a program to create a binary search tree and perform different operations on it.

The output of the program is as follows:

```
>>> root=insert(root,10)
>>> root=insert(root,7)
>>> root=insert(root,5)
>>> root=insert(root,8)
>>> root=insert(root,15)
>>> root=insert(root,11)
>>> root=insert(root,18)
>>> printInorder(root)
5
7
8
10
11
15
18
```

```
>>> printPreorder(root)
10
7
5
8
15
11
18
>>> printPostorder(root)
5
8
7
11
18
15
10
>>> root=deleteIterative(root,15)
>>> printInorder(root)
5
7
8
10
11
18
```

8.4.3 Creating a Binary Tree Using Traversal Methods

A binary tree can be constructed if we are given at least two of the traversal results, provided that one traversal is always an in-order traversal and the second is either a pre-order traversal or a post-order traversal. An in-order traversal determines the left and right child nodes of the binary tree. A pre-order or post-order traversal determines the root node of the binary tree. Hence, there are two different ways of creating a binary tree:

1. In-order and pre-order traversal

2. In-order and post-order traversal

Now, we have pre-order and in-order traversal sequences. Then, the following steps are followed to construct a binary tree:

Step 1: The pre-order traversing sequence is used to determine the root node of the binary tree. The first node in the pre-order sequence is the root node.

Step 2: The in-order traversing sequence is used to determine the left and the right sub-trees of the binary tree. Keys toward the left side of the root node in the in-order sequence form the left sub-tree. Similarly, keys toward the right side of the root node in the in-order sequence form the right sub-tree.

Step 3: Each element from the pre-order traversing sequence is recursively selected and the left and the right sub-trees are created from the in-order traversing sequence.

For Example – Create a binary tree from the given traversing sequences.

In-order – A C D E I N O T U
Pre-order – E DCAU T I O N

Now, we construct the binary tree.

1. The first node in the pre-order sequence is the root node of the tree. Hence, E is the root node of the binary tree.

$$\boxed{E}$$

Root Node

2. Now, we can easily determine the left and right sub-trees from the in-order sequence. Keys toward the left side of the root node, that is, A, C, and D, form the left sub-tree. Similarly, elements on the right side of the root node, that is, I, N, O, T, and U, form the right sub-tree.

$$\boxed{E}$$

Root Node

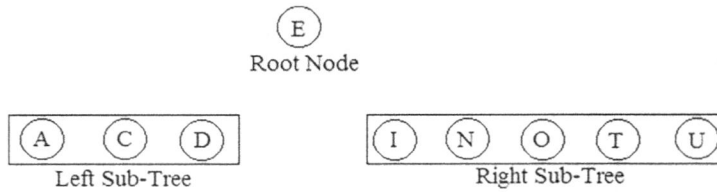

Root Node

Left Sub-Tree   Right Sub-Tree

3. Now, the left child of the root node is the first node in the pre-order traversing sequence after the root node E. Thus, D is the left child of the root node E.

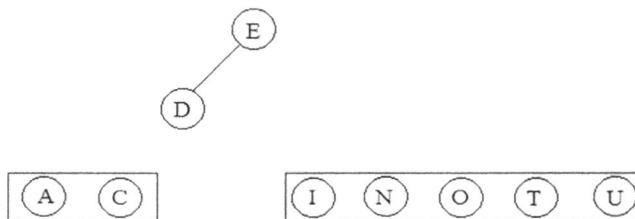

4. Similarly, the right child of the root node is the first node in the pre-order traversing sequence after the nodes of the left sub-tree. Thus, U is the right child of the root node E.

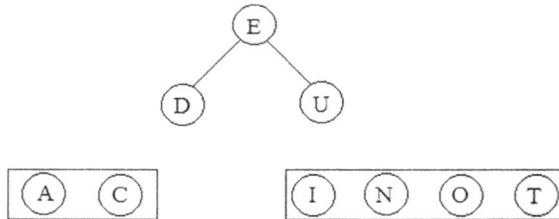

5. In the in-order sequence, A and C are on the left side of D. So, A and C form the left sub-tree of D.

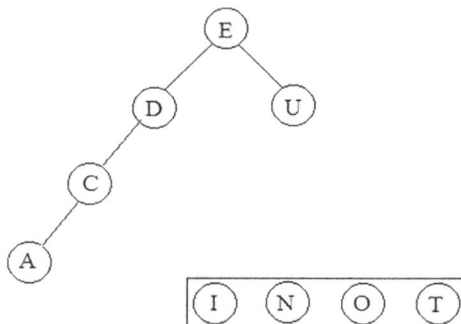

6. The next elements in the pre-order sequence are T and I. In the in-order sequence, T and I are on the left side of U. So, T and I form the left sub-tree of U.

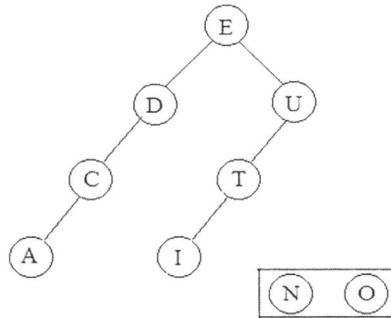

7. The next element in the pre-order sequence is O. In the in-order sequence, O is on the right side of I. So, O forms the right sub-tree of I. The last element in the pre-order sequence is N. N is on the left side of O in the in-order sequence. Thus, N forms the left sub-tree of O.

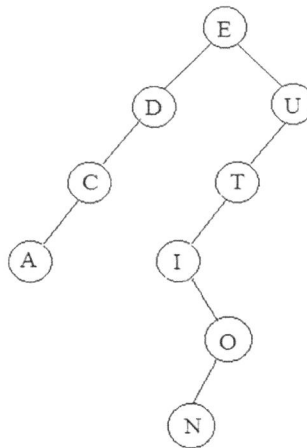

Finally, the binary tree is created from the given traversing sequences.

Frequently Asked Questions

Q. Create a binary tree from the given traversing sequences.

In-order – d b e a f c g

Pre-order – a b d e c f g

Answer:

Step 1: a is the root node of the binary tree.

Step 2: d, b, and e are on the left side of the a node in the in-order sequence. Hence, d, b, and e are the left sub-trees of root a. Node d is the left sub-tree of b, and e is the right sub-tree of b.

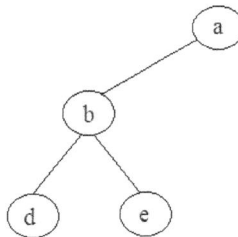

Step 3: f, c, and g are on the right side of root a in the in-order sequence. Hence, f, c, and g are the right sub-trees of root a. Node f is the left sub-tree of c and g is the right sub-tree of c.

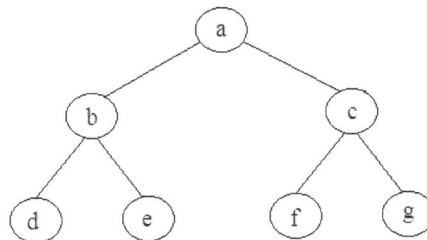

8.5 AVL TREES

The AVL tree was invented by Adelson-Velski and Landis in 1962. The AVL tree is so named in honor of its inventors. The AVL tree was the first balanced binary search tree. It is a self-balancing binary search tree. The AVL tree is also known as a *height-balanced tree* because of its property that the heights of the two sub-trees of a node can differ at most by one. AVL trees are very efficient in performing searching, insertion, and deletion operations, as they take O(log n) time to perform all these operations.

8.5.1 Need for Height-Balanced Trees

AVL trees are very similar to binary search trees but with a small difference. AVL trees have a special variable known as a balance factor associated with them. Every node in the AVL tree has a balance factor associated with it. The balance factor is determined by subtracting the height of the right sub-tree from the height of the left sub-tree. Thus, a node with a balance factor of –1, 0, or 1 is said to be a height-balanced tree. The primary need for the height-balanced tree is that the process of searching becomes very fast. This balancing condition also ensures that the depth of the tree is O(logn). The balance factor is calculated as follows:

Balance Factor = Height(Left sub-tree) – Height(Right sub-tree)

- If the balance factor of the tree is –1, it means that the height of the right sub-tree of that node is one more than the height of the left sub-tree of that node.

- If the balance factor of the tree is 0, it means that the height of the left and the right sub-trees of a node are equal.

- If the balance factor of the tree is 1, it means that the height of the left sub-tree of that node is one more than the height of its right sub-tree.

Thus, the overall benefit of the height-balanced tree is to assist in fast searching.

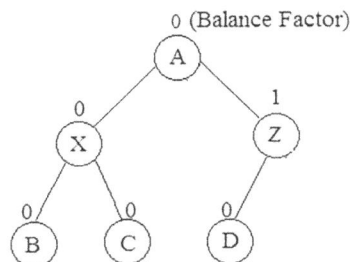

FIGURE 8.21 A balanced AVL tree

8.5.2 Operations on an AVL Tree

In this section, we discuss various operations that are performed on AVL trees. These are as follows:

- Searching a node in an AVL Tree
- Inserting a new node in an AVL Tree

1. Searching a node in an AVL Tree

The process of searching a node in an AVL tree is the same as for a binary search tree.

2. Inserting a new node in an AVL Tree

The process of inserting a new node in an AVL tree is quite similar to that of binary search trees. The new node is always inserted as a terminal/leaf node in the AVL tree. But the insertion of a new node can disturb the balance of the AVL tree, as the balance factor may be disturbed. Thus, for the tree to remain balanced, the insertion process is followed by a rotation process. The rotation process is usually done to restore the balance factor of the tree. If the balance factor of each node is –1, 0, or 1 after the insertion process, then the rotation is not required, as the tree is already balanced; otherwise, rotation is required. Now, let us look at the given example and see how insertion is done without rotations.

For Example – In the given AVL tree, insert a new node with value 60 in the tree.

Initially, the AVL tree is given as

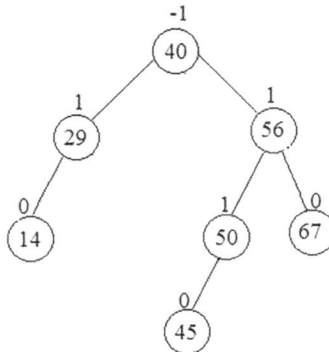

FIGURE 8.22 The AVL tree before insertion

We insert 60 into the AVL tree.

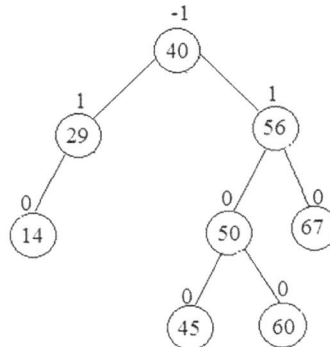

FIGURE 8.23 The AVL tree after inserting 60

Hence, after insertion, there are no nodes in the tree that are unbalanced. Thus, there is no need to apply rotation here. However, now we discuss how the rotation process is performed in AVL trees.

AVL Rotations

Rotation is done when the balance factor of the node becomes disturbed after inserting a new node. We know that the new node that is inserted will always have a balance factor of 0, as it is a leaf node. Hence, the nodes whose balance factors will be disturbed are the ones that lie in the path of the root node to the newly inserted node. So, we perform the rotation process only on those nodes whose balance factors will be disturbed. In the rotation process, our first work is to find the critical node in the AVL tree. The critical node is the nearest ancestor node from the newly inserted node to the root node that does not have a balance factor of –1, 0, or 1. First, let us understand the concept of the critical node with the help of an example.

For Example – Find the critical node in the given AVL tree.

Initially, the AVL tree is given as follows:

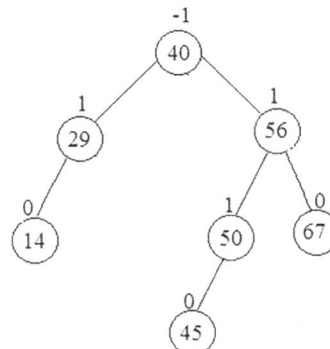

FIGURE 8.24 An AVL tree

We insert a new node with value 42 in the tree.

After inserting 42 in the AVL tree, we can see that there are three nodes in the tree that have balance factors equal to –2, 2, and 2. Now, the critical node is the one that is the nearest to the newly inserted node with a disturbed balance factor. We can see that 50 is the nearest node to 42, and 50 has a balance factor of 2. Thus, 50 is the critical node in this AVL tree. However, to restore the balance factor of the previous AVL tree, rotations are performed. There are four types of rotations:

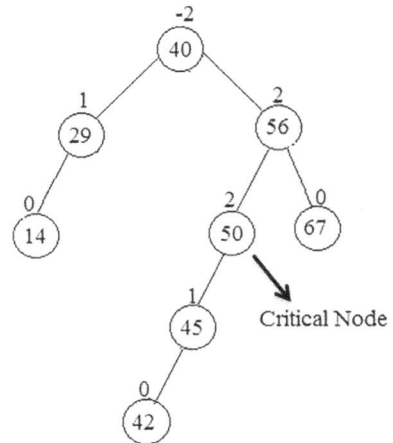

FIGURE 8.25 The AVL tree

1. **Left-Left Rotation (LL Rotation)** – A new node is inserted in the left sub-tree of the left sub-tree of the critical node.

2. **Right-Right Rotation (RR Rotation)** – A new node is inserted in the right sub-tree of the right sub-tree of the critical node.

3. **Right-Left Rotation (RL Rotation)** – A new node is inserted in the left sub-tree of the right sub-tree of the critical node.

4. **Left-Right Rotation (LR Rotation)** – A new node is inserted in the right sub-tree of the left sub-tree of critical node.

LL Rotation

The LL rotation is also known as the Left-Left rotation, as the new node is inserted in the left sub-tree of the left sub-tree of the critical node. It is a single rotation. Let us take an example and perform an LL rotation in it.

For Example –

Initially, the AVL tree is given as

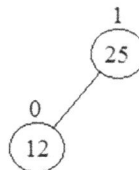

FIGURE 8.26(a)

Insert new node 5 in the AVL tree.

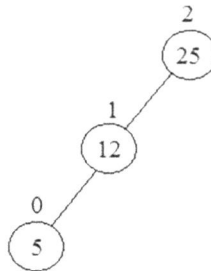

FIGURE 8.26(b)

After inserting 5 in the AVL tree, the balance factor of 25 is disturbed. Thus, 25 is the critical node. Hence, we apply the LL rotation to restore the balance factor of the tree. After rotation node 12 becomes the root node, node 5 and node 25 become the left and the right child of the tree, respectively.

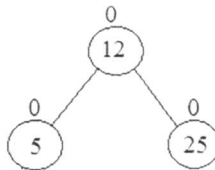

FIGURE 8.26 Showing an LL rotation in an AVL tree

Therefore, the LL rotation is performed, and the balance factor of each node is also restored.

RR Rotation

The RR rotation is also known as a Right-Right rotation, as the new node is inserted in the right sub-tree of the right sub-tree of the critical node. It is also a single rotation. Let us take an example and perform an RR rotation in it.

For Example –

Initially the AVL tree is given as follows:

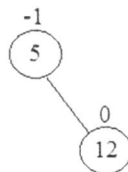

FIGURE 8.27(a)

Insert new node 25 in the AVL tree.

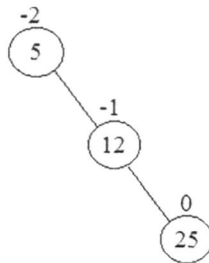

FIGURE 8.27(b)

After inserting 25 in the AVL tree, the balance factor of 5 is disturbed. Thus, 5 is the critical node. Hence, here we apply an RR rotation to restore the balance factor of the tree. After rotation node 12 becomes the root node, node 5 and node 25 become the left and the right child of the tree, respectively.

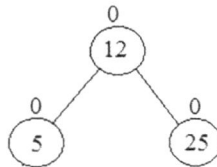

FIGURE 8.27 Showing an RR rotation in an AVL tree

Therefore, the RR rotation is performed, and the balance factor of each node is also restored.

RL Rotation

The RL rotation is also known as a Right-Left rotation, as the new node is inserted in the left sub-tree of the right sub-tree of the critical node. It is a double rotation. Let us take an example and perform an RL rotation in it.

For Example –
Initially, the AVL tree is given as follows:

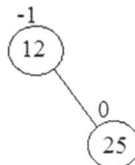

FIGURE 8.28(a)

Insert new node 15 in the AVL tree.

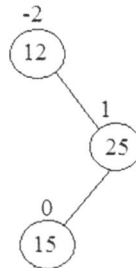

FIGURE 8.28(b)

After inserting 15 in the AVL tree, the balance factor of 12 is disturbed. Thus, 12 is the critical node. Hence, here we apply an RL rotation to restore the balance factor of the tree. After rotation node 15 becomes the root node, node 12 and node 25 become the left and the right child of the tree, respectively.

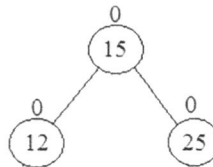

FIGURE 8.28 Showing an RL rotation in an AVL tree

Therefore, the RL rotation is performed, and the balance factor of each node is also restored.

LR Rotation

The LR rotation is also known as a Left-Right rotation, as the new node is inserted in the right sub-tree of the left sub-tree of the critical node. It is also a double rotation. Let us take an example and perform an LR rotation in it.

For Example –

Initially, the AVL tree is given as follows:

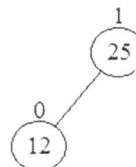

FIGURE 8.29(a)

Insert new node 15 in the AVL tree.

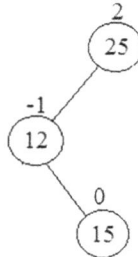

FIGURE 8.29(b)

After inserting 15 in the AVL tree, the balance factor of 25 is disturbed. Thus, 25 is the critical node. Hence, here we apply an LR rotation to restore the balance factor of the tree. After rotation node 15 becomes the root node, and node 12 and node 25 become the left and the right child of the tree, respectively.

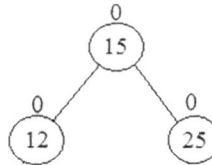

FIGURE 8.29 Showing an LR rotation in an AVL tree

Therefore, an LR rotation is performed, and the balance factor of each node is also restored.

Frequently Asked Questions

Q. Create an AVL tree by inserting the following elements.

60, 10, 20, 30, 19, 120, 100, 80, 19

Answer:

Step 1: Insert 60.

Step 2: Insert 10. No rebalancing is required.

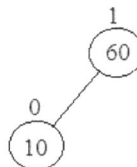

Step 3: *Insert 20. Now, rebalancing is required. We perform the LR rotation.*

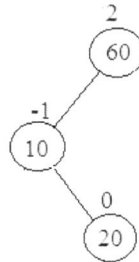

Step 4: *After performing the LR rotation, the AVL tree is given as*

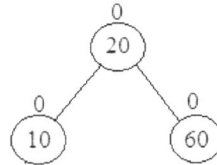

Step 5: *Insert 30. No rebalancing is required.*

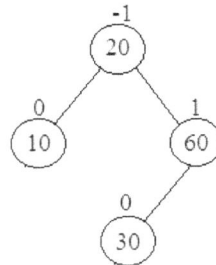

Step 6: *Insert 19. No rebalancing is required.*

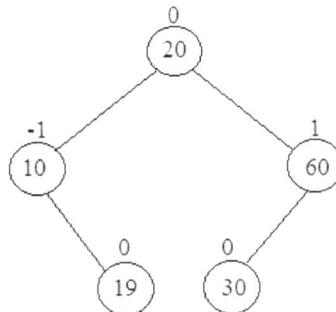

Step 7: *Insert 120. No rebalancing is required.*

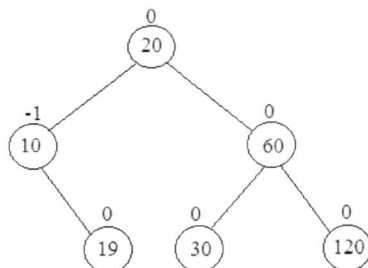

Step 8: *Insert 100. No rebalancing is required.*

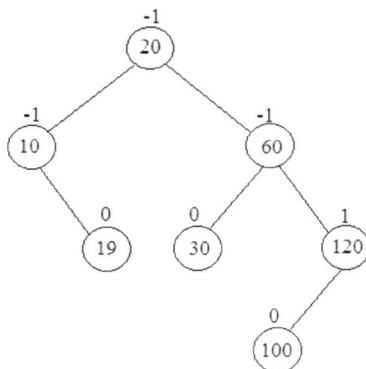

Step 9: *Insert 80. Now, rebalancing is required. We perform the LL rotation.*

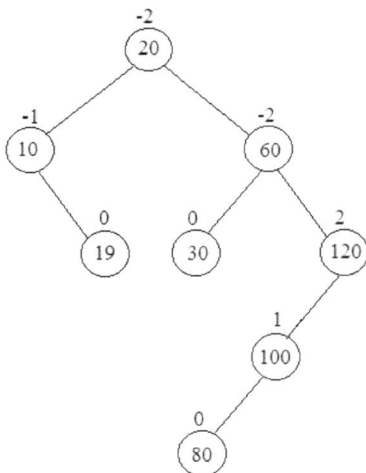

Step 10: *After performing the LL rotation, the AVL tree is given as*

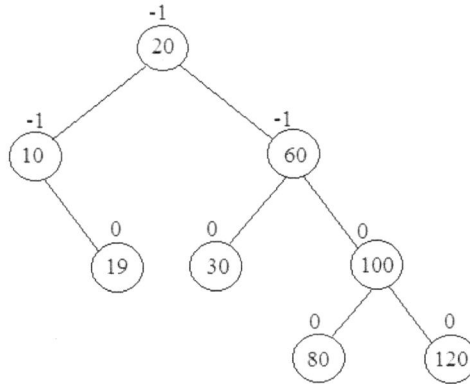

Step 11: *Insert 19. Now, rebalancing is required. We perform the RR rotation.*

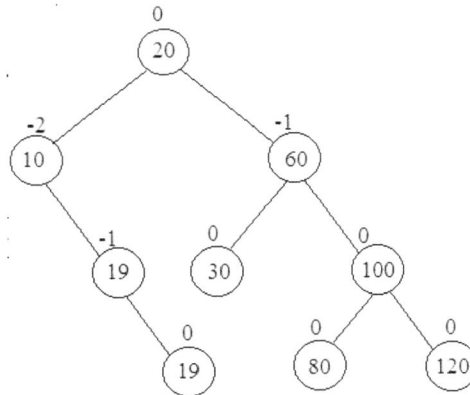

Step 12: *After performing RR rotation, the AVL tree becomes*

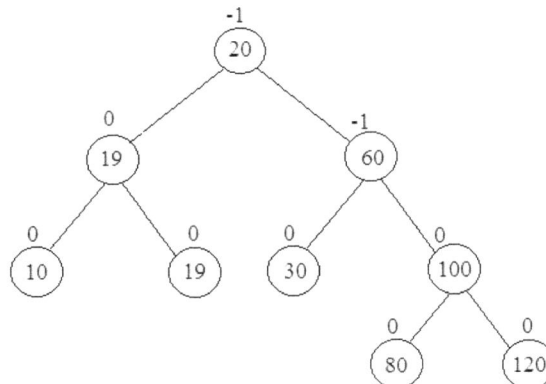

8.6 SUMMARY

- A tree is defined as a collection of one or more nodes where one node is designated as a root node, and the remaining nodes can be partitioned into the left and the right sub-trees. It is used to store hierarchical data.

- The root node is the topmost node of the tree. It does not have a parent node. If the root node is empty, then the tree is empty. A leaf node is one that does not have any child nodes.

- A path is a unique sequence of consecutive edges that is required to be followed to reach the destination from a given source.

- The degree of a node is equal to the number of children that a node has.

- A binary tree is a collection of nodes where each node contains three parts, that is, a left child, a right child, and the data item. A binary tree can have at most 2 children; that is, a parent can have either 0, 1, or 2 children.

- There are two types of binary trees, that is, complete binary trees and extended binary trees.

- In a complete binary tree, every level except the last one must be completely filled. All the nodes in the complete binary tree must appear left as much as possible.

- Extended binary trees are also known as 2T-trees. A binary tree is said to be an extended binary tree if and only if every node in the binary tree has either 0 children or 2 children.

- Binary trees can be represented in the memory in two ways, as an array representation of binary trees and a linked representation of binary trees. Array representation, also known as sequential representation, of binary trees is done using one-dimensional (1D) arrays. The linked representation of binary trees is done using linked lists.

- A Binary Search Tree (BST) is a variant of a binary tree in that all the nodes in the left sub-tree have a value less than that of a root node. Similarly, all the nodes in the right sub-tree have a value more than that of a root node. It is also known as an ordered binary tree.

- The search operation is one of the most common operations performed in a binary search tree. This operation is performed to find whether a particular key exists in the tree.

- An insertion operation is performed to insert a new node with the given value in a binary search tree.

- The mirror image of a binary search tree means interchanging the right sub-tree with the left sub-tree at every node of the tree.

- Traversing is the process of visiting each node in the tree exactly once in a particular order. A tree can be traversed in various ways, via pre-order traversal, in-order traversal, and post-order traversal.

- The word "pre" in "pre-order" determines that the root node is accessed before accessing any other node in the tree. Hence, it is also known as a DLR traversal, that is, Data Left Right.
- The word "in" in "in-order" determines that the root node is accessed in between the left and the right sub-trees. Hence, it is also known as an LDR traversal, that is, Left Data Right.
- The word "post" in "post-order" determines that the root node will be accessed last after the left and the right sub-trees. Hence, it is also known as an LRD traversal, that is, Left Right Data.
- A binary tree can be constructed if we are given at least two of the traversal results, provided that one traversal should always be an in-order traversal and the second can be either a pre-order traversal or post-order traversal.
- An AVL is a self-balancing binary search tree. Every node in the AVL tree has a balance factor associated with it. The balance factor is calculated by subtracting the height of the right sub-tree from the height of the left sub-tree. Thus, a node with a balance factor of -1, 0, or 1 is said to be a height-balanced tree.

8.7 EXERCISES

8.7.1 Theory Questions

Q1. What is a tree? Discuss its various applications.

Q2. Differentiate between height and level in a tree.

Q3. Explain the concept of binary trees.

Q4. In what ways can a binary tree be represented in the computer's memory?

Q5. What is meant by a binary search tree?

Q6. List the various operations performed on binary search trees.

Q7. How can a node be deleted from a binary search tree? Discuss all the cases in detail with examples.

Q8. Create a binary search tree by inserting the following keys – 76, 12, 56, 31, 199, 17, 40, 76, 75. Also, find the height of the binary search tree.

Q9. Create a binary search tree by performing following operations:
   (i.) Insert 50, 34, 23, 87, 100, 67, 43, 51, 18, and 95.
   (ii.) Delete 100, 34 and 95, 50 from the binary search tree.
   (iii.) Find the smallest value in the binary search tree.

Q10. How can we find the mirror image of a binary search tree?

Q11. List the various traversal methods of a binary tree.

Q12. What do you understand about an AVL tree?

Q13. Explain the concept of the balance factor in AVL trees.

Q14. List the advantages of an AVL tree.

Q15. Consider the following binary search tree and perform the following operations:

(i.) Find the pre-order and post-order traversals of the tree.

(ii.) Insert 25, 32, 50, 75, and 87 in the tree.

(iii.) Find the largest value in the tree.

(iv.) Delete the root node.

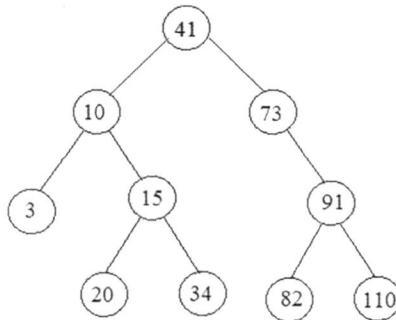

Q16. Give the linked representation of the given binary search tree.

Q17. Construct a binary search tree of the word VIVEKANANDA. Find its pre-order, in-order, and post-order traversal.

Q18. Create an AVL tree by inserting the following keys, 50, 19, 59, 90, 100, 12, 10, and 150, into the tree.

Q19. Consider the following AVL search tree and perform various operations in it:

(i.) Insert 100, 58, 93, 40, and 7 into the tree.

(ii.) Search for 93 in the AVL tree.

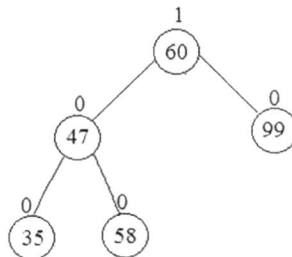

Q20. Discuss the various types of rotations performed in AVL trees.

Q21. Which of the following is better and why?

    (i.)   AVL trees

    (ii.)   Binary search trees

Q22. Consider the following tree and identify the following aspects of it:

    (i.)   Determine the height of the tree.

    (ii.)   Name the leaf nodes.

    (iii.) Siblings of C.

    (iv.)   Level number of the node J.

    (v.)   Root node of the tree.

    (vi.)   Left and right sub-trees.

    (vii.) Depth of the tree.

    (viii.) Ancestors of E.

    (ix.)   Descendants of H.

    (x.)   Path from node A to F.

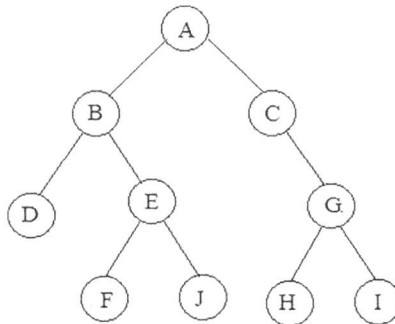

8.7.2 Programming Questions

  Q1. Write a function to find the height of a binary search tree.

  Q2. Write a Python program to insert and delete nodes from a binary search tree.

  Q3. Write a Python program to show insertion in AVL trees using classes.

  Q4. Write a function to calculate the total number of nodes in a tree.

  Q5. Write a Python program to traverse a binary search tree showing all the traversal methods using classes.

Q6. Write a function to find the largest value in a binary search tree.

Q7. Write an algorithm showing the post-order traversal of a binary search tree.

Q8. Write an algorithm to find the total number of internal nodes in a binary search tree.

Q9. Write a function to search for a node in a binary search tree.

8.7.3 Multiple Choice Questions

Q1. The maximum height of a binary tree with n number of nodes is _____.

 a. 0

 b. n

 c. n+1

 d. n−1

Q2. The degree of a terminal node is always _____.

 a. 1

 b. 2

 c. 0

 d. 3

Q3. A binary tree is a tree in that _____.

 a. Every node must have two children

 b. Every node must have at least two children

 c. No node can have more than two children

 d. All of these

Q4. What is the post-order traversal of the binary search tree having pre-order traversal as DBAEFGCH and in-order traversal as BEAFDCHG?

 a. EFBAHGCD

 b. EFBAHCGD

 c. EFABHGCD

 d. EFABHCGD

Q5. How many rotations are required during the construction of an AVL tree if the following keys are to be added in the order given?
36, 51, 39, 24, 29, 60, 79, 20, 28

   a. 3 Left rotations, 3 Right rotations

   b. 2 Left rotations, 2 Right rotations

   c. 2 Left rotations, 3 Right rotations

   d. 3 Left rotations, 2 Right rotations

Q6. A binary tree of height h has at least h nodes and at most _____ nodes.

   a. 2

   b. 2^h

   c. $2^h - 1$

   d. $2^h + 1$

Q7. How many distinct binary search trees can be created out of four distinct keys?

   a. 5

   b. 12

   c. 14

   d. 23

Q8. Nodes at the same level that also share same parent are called _____.

   a. Cousins

   b. Siblings

   c. Ancestors

   d. Descendants

Q9. The balance factor of a node is calculated by _____.

   a. $\text{Height}_{\text{Left sub-tree}} - \text{Height}_{\text{Right sub-tree}}$

   b. $\text{Height}_{\text{Right sub-tree}} - \text{Height}_{\text{Left sub-tree}}$

   c. $\text{Height}_{\text{Left sub-tree}} + \text{Height}_{\text{Right sub-tree}}$

   d. $\text{Height}_{\text{Right sub-tree}} + \text{Height}_{\text{Left sub-tree}}$

Q10. The following sequence is inserted into an empty binary search tree:
6 11 26 12 5 7 16 8 35
What is the type of traversal given by the following numbers?
6 5 11 7 26 8 12 35 16

    a. Pre-order traversal   **b.** In-order traversal

    c. Post-order traversal   **d.** None of these

Q11. In tree creation, which one will be the most suitable and effective data structure?

    a. Stack   **b.** Linked list

    c. Queue   **d.** Array

Q12. A binary tree can be represented as

    a. Linked List   **b.** Arrays

    c. Both of the above   **d.** None of the above

Q13. A binary tree of n nodes has exactly n+1 edges.

    a. True

    b. False

    c. Not possible to comment

Q14. The in-order traversal of a tree will yield a sorted listing of the elements of trees in

    a. Binary heaps   **b.** Binary trees

    c. Binary search trees   **d.** All of these

Q15. What is the nearest ancestor node on the path from the root node to the newly inserted node of the AVL tree having balance factor -1, 0, or 1?

    a. Parent node   **b.** Child node

    c. Root node   **d.** Critical node

## CHAPTER 9

MULTI-WAY SEARCH TREES

9.1 INTRODUCTION

We studied binary search trees and discussed that every node in a binary search tree contains three parts: an information part, LEFT child, and RIGHT child that stores the address to the left and right sub-trees. The same concept is used for multi-way search trees. An *M-way search tree* is a tree that contains (M − 1) values per node. It also has M sub-trees. In an M-way search tree, M is called the degree of the node. For example, if the value of M = 3 in an M-way search tree, then the tree contains two values per node and it has three sub-trees. When an M-way search tree is not empty, it has the following properties:

1. Each node in an M-way search tree is of the following structure:

n	P_0	K_0	P_1	K_1	P_2	K_2 _ _ _ _ _ _	P_{n-1}	K_{n-1}	P_n

   where P_0, P_1, P_2, . . . P_n are the node's sub-trees, and K_0, K_1, K_2, . . . K_n are the key values stored in the node.

2. The key values in a node are stored in ascending order, that is, $K_i < K_{i+1}$, where i = 0, 1, 2, . . . n−2.

3. All the key values stored in the left subtree are always less than the root node.

4. All the key values stored in the right subtree are always greater than the root node.

5. The sub-trees pointed to by P_i for i = 0, 1, 2, . . .n are also M-way search trees.

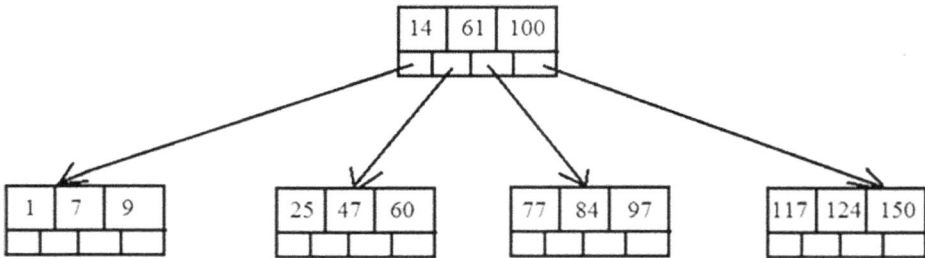

FIGURE 9.1 An M-way search tree of order 4

9.2 B-TREES

A *B-tree* is a specialized multi-way tree that is widely used for disk access. The B-tree was developed in 1970 by Rudolf Bayer and Ed McCreight. In a B-tree, each node may contain a large number of keys. A B-tree is designed to store a large number of keys in a single node so that the height remains relatively small. A B-tree of order m has all the properties of a multi-way search tree. In addition, it has the following properties:

1. All leaf nodes are at the bottom level or at the same level.

2. Every node in a B-tree can have at most m children.

3. The root node can have at least two children if it is not a leaf node, and it can obviously have no children if it is a leaf node.

4. Each node in a B-tree can have at least (m/2) children except the root node and the leaf node.

5. Each leaf node must contain at least ceil $[(m/2) - 1]$ keys.

For example – A B-tree of order 5 can have at least ceil $[5/2] = 3$ children and ceil $[(5/2) - 1] = 2$ keys. Obviously, the maximum number of children a node can have is 5. Each leaf node must contain at least 2 keys.

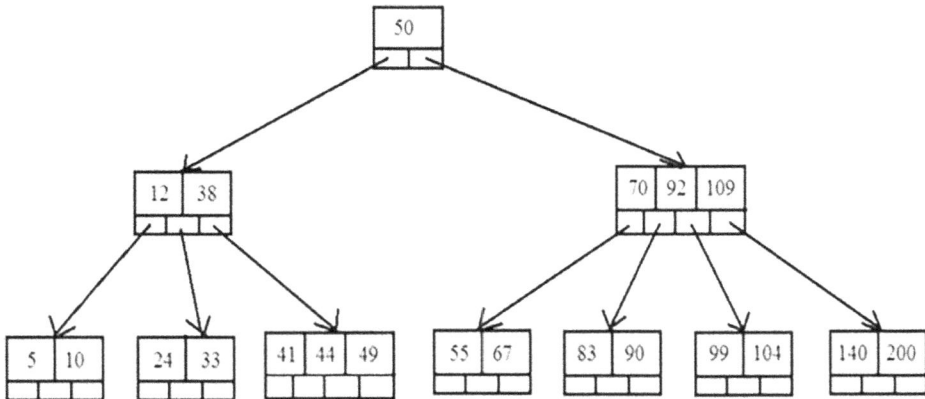

FIGURE 9.2 A B-tree of order 4

> **Practical Application:**
>
> In database programs, the data is too large to fit in memory; therefore, it is stored in the secondary storage, that is, tapes or disks.

9.3 OPERATIONS ON A B-TREE

B-tree stores sorted data, and we can perform on it the following operations:

- Inserting a new element in a B-tree
- Deleting an element from a B-tree

9.3.1 Insertion in a B-Tree

Insertions in a B-tree are done at the leaf-node level. The following are the steps for inserting an element in a B-tree:

Step 1 – In Step 1, we will search the B-tree to find the leaf node where the new key is to be inserted.

Step 2 – Now, if the leaf node is full, that is, if it already contains (m – 1) keys, then follow these steps:

i. Insert the new key into the existing set of keys in order.

ii. Now, the node is split into two halves.

iii. Finally, push the middle (median) element upward to its parent node. Also, if the parent node is full, then split the parent node by following these steps.

Step 3 – If the leaf node is not full, that is, if it contains (m – 1) keys, then insert the new key into the node, keeping the elements of the node in order.

Frequently Asked Questions

Q. Construct a B-tree of order 5 and insert the following values into it:

Values to be inserted – B, N, G, A, H, E, J, Q, M, D, V, L, T, Z

Answer:

1. *Since order = 5, we can store at least 3 values and at most 4 values in a single node. Hence, we insert B, N, G, and A into the B-tree in sorted order.*

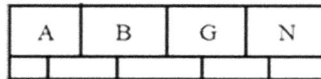

A	B	G	N

FIGURE 9.3(a)

2. *H is inserted between G and N, so the order is A B G H N. That is not possible, as at most 4 values can be accommodated in a single node. So now we will split the node, and the middle element G becomes the root node.*

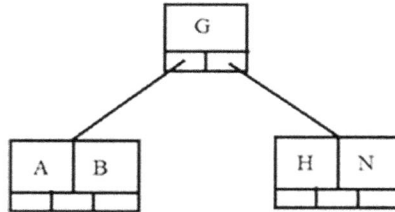

FIGURE 9.3(b)

3. *Now we insert E J and Q into the B-tree.*

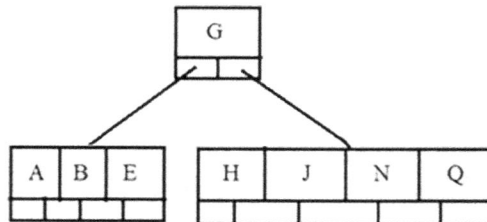

FIGURE 9.3(c)

4. *M is to be inserted in the right subtree. But at most 4 values can be stored in the node, so now we push the middle element, that is, M, into the root node. Thus, the node is split into two halves.*

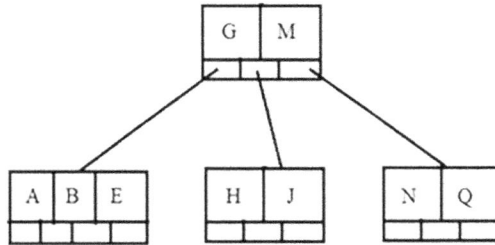

FIGURE 9.3(d)

5. *We insert D V L and T into the tree.*

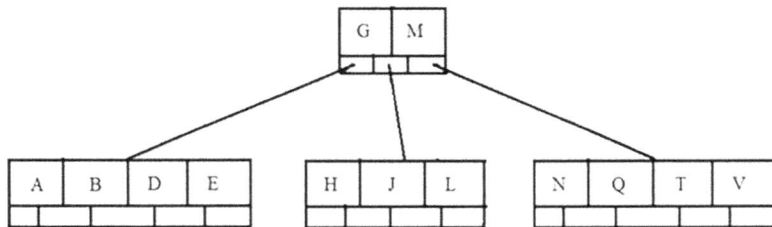

FIGURE 9.3(e)

6. *Finally, Z is inserted. It is inserted in the right subtree. Hence, the last node is split into two halves, and the middle element, that is, T, pushes up to the root node.*

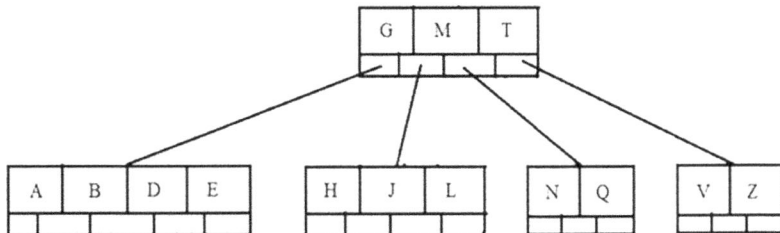

FIGURE 9.3(f)

9.3.2 Deletion in a B-Tree

The deletion of keys in a B-tree requires traversal in the B-tree; that is, after reaching a particular node, we can come across two cases:

1. The node is a leaf node.

2. The node is not a leaf node.

1. The node is a leaf node.

If the node has more than a minimum number of keys, then deletion can be done very easily. But if the node has a minimum number of keys, then first we check the number of keys in the adjacent leaf node. If the number of keys in the adjacent node is greater than the minimum number of keys, then the first key of the adjacent leaf node goes to the parent node and the key present in the parent node is combined in a single leaf node. If the parent node also has less than the minimum number of keys, then the same steps are repeated until we get a node that has more than the minimum number of keys present in it.

2. The node is not a leaf node.

In this case, the key from the node is deleted, and its place is occupied by either its successor or predecessor key. If both predecessor and successor nodes have keys less than the minimum number, then the keys of the successor and predecessor are combined.

For Example – Consider a B-tree of order 5.

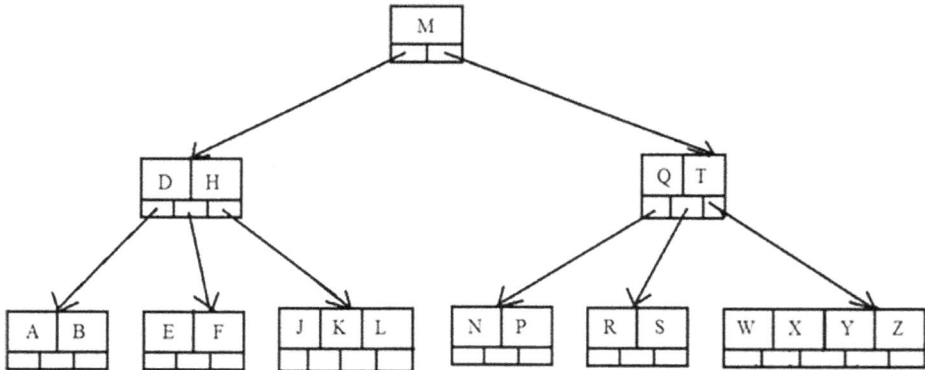

FIGURE 9.4(a)

1. Delete J from the tree. J is in the leaf node, so it is simply deleted from the B-tree.

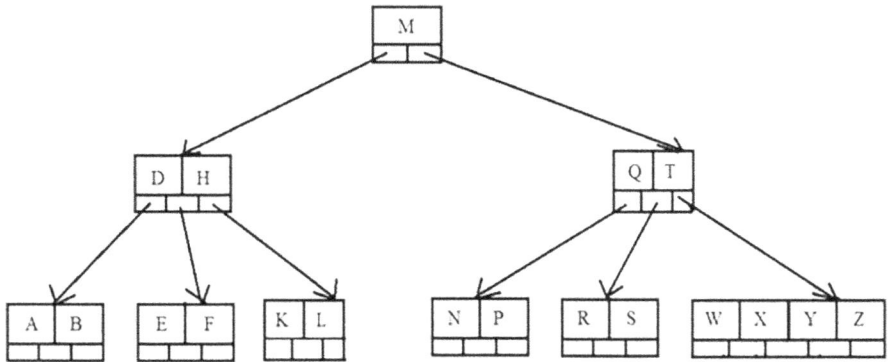

FIGURE 9.4(b)

2. Now T is to be deleted, but it is not in the leaf node, so we replace T with its successor, that is, W. Hence, T is deleted.

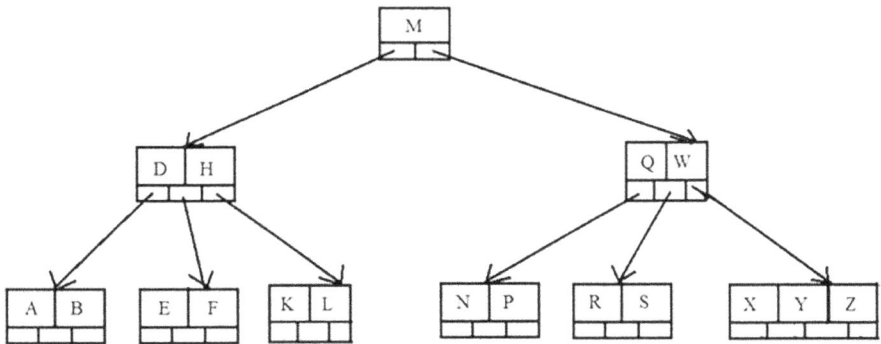

FIGURE 9.4(c)

3. Now delete R; in this case, we borrow keys from the adjacent leaf node.

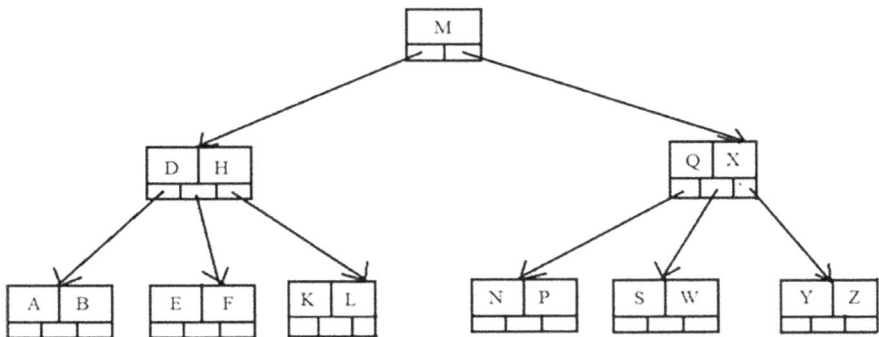

FIGURE 9.4(d)

4. Now we want to delete E. In this case, we also borrow keys from an adjacent node. But we can see that there are no free keys in an adjacent node, so the leaf node has to be combined with one of its two siblings. This includes moving down the parent's key that was between those two leaves.

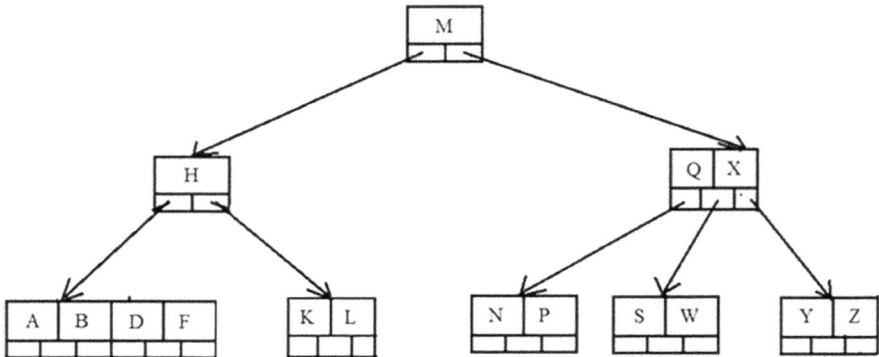

FIGURE 9.4(e)

But we can see that H is still unstable according to the definition. Therefore, the final tree after all deletions is as follows:

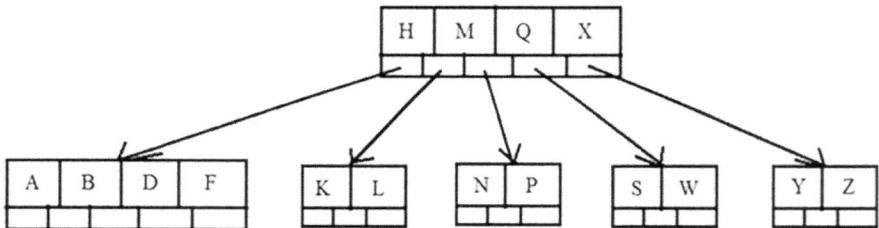

FIGURE 9.4(f)

Frequently Asked Questions

Q. Consider the following B-tree of order 5 and insert 81, 7, 49, 61, and 30 into it.

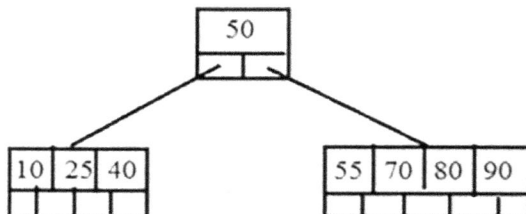

Answer:

1. *Insert 81.*

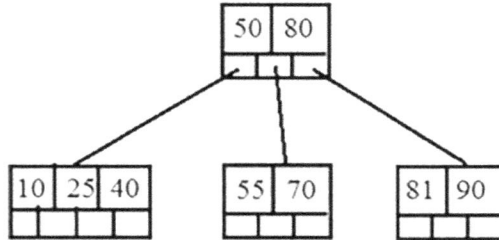

2. *Insert 7 and 49.*

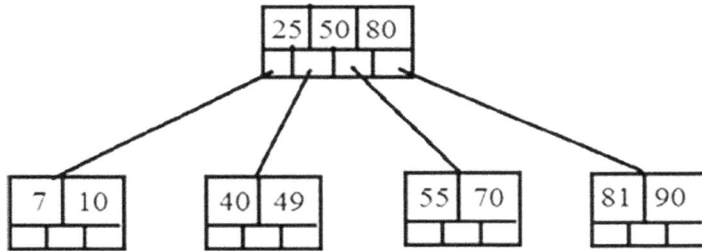

3. *Insert 61 and 30.*

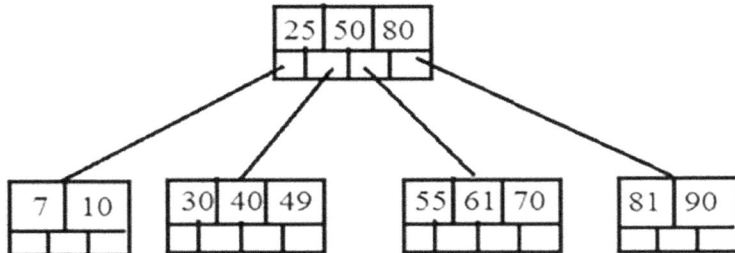

FIGURE 9.5 Insertion in a B-tree

Frequently Asked Questions

Q. Consider the following B-tree of order 5 and delete the values 95, 200, 176, and 70 from it.

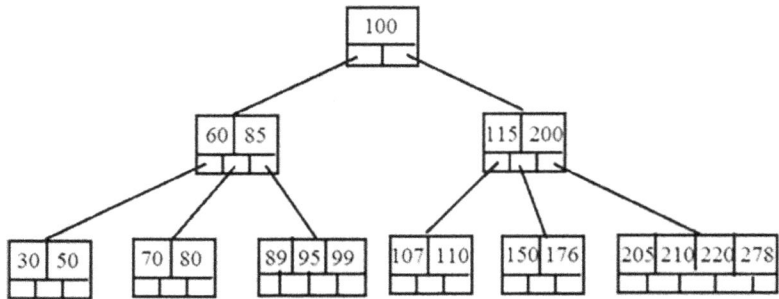

Answer:

1. *Delete 95.*

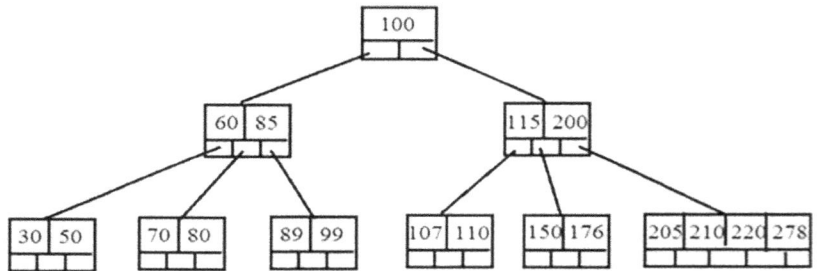

2. *Delete 200.*

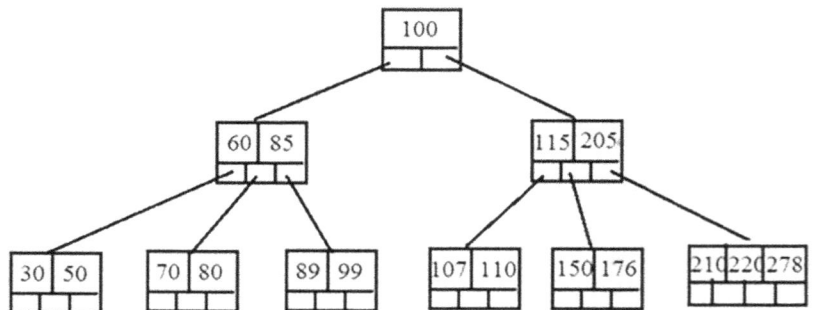

3. *Delete 176.*

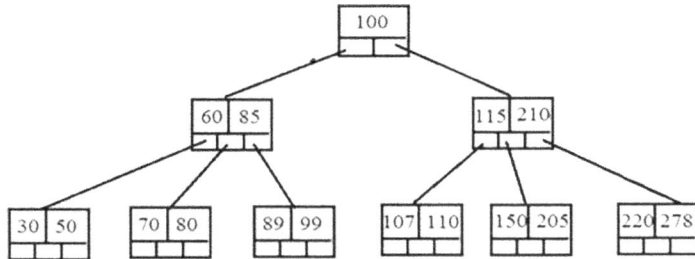

4. *Delete 70.*

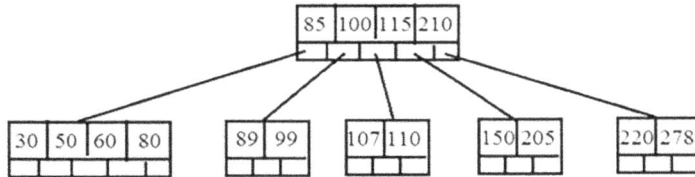

FIGURE 9.6 Deletion in a B-tree

9.4 APPLICATION OF A B-TREE

The main application of a B-tree is the organization of a large amount of data or a huge collection of records into a file structure. A B-tree should search the records very efficiently, and all the operations, such as insertion, deletion, and searching, should be done very efficiently; therefore, the organization of records should be very good.

9.5 B+ TREES

A *B+ tree* is a variant of a B-tree that also stores sorted data like a B-tree. The structure of a B-tree is the standard organization for indexes in database systems. Multilevel indexing is done in a B+ tree; that is, leaf nodes constitute a dense index, while non-leaf nodes constitute a sparse index. A B+ tree is a slightly different data structure that allows sequential processing of data and stores all the data in the lowest level of the tree. A B-tree can store both records and keys in its interior nodes, while a B+ tree stores all the records in its leaf nodes and the keys in its interior nodes. In a B+ tree, the leaf nodes are linked to one another like a linked list. A B+ tree is usually used to store large amounts of data that cannot be stored in the primary memory. Hence, in a B+ tree, the leaf nodes are stored in the secondary storage, while the internal nodes are stored in the main memory.

In a B+ tree, all the internal nodes are called *index nodes* because they store the index values. Similarly, all the external nodes are called *data nodes* because they store the keys. A B+ tree is always balanced and is very efficient for searching data, as all the data is stored in the leaf nodes. The advantages of a B+ tree are as follows:

a. A B+ tree is always balanced, and the height of the tree always remains low.

b. All the leaf nodes are linked to one another, which makes it very efficient.

c. The leaf nodes are also linked to the nodes at an upper level; thus, it can be easily used for a wide range of search queries.

d. The records can be fetched in an equal number of disk access.

e. The records can be accessed either sequentially or randomly.

f. Searching data is very simple, as all the information is stored only in the leaf nodes.

g. Similarly, deletion is also very simple, as it only takes place in the leaf nodes.

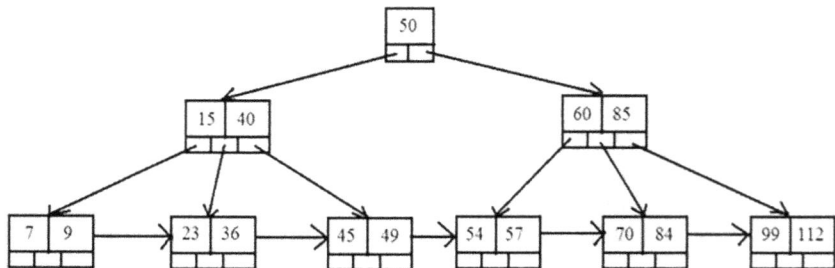

FIGURE 9.7 B+ tree of order 3

9.6 SUMMARY

- An M-way search tree has M − 1 values per node and M sub-trees. M is called the degree of the node.
- A B-tree is a specialized multi-way tree that is widely used for disk access. The B-tree was developed in 1970 by Rudolf Bayer and Ed McCreight.

- A B-tree of order m has all the properties of a multi-way search tree.
- The main application of a B-tree is the organization of a large amount of data or a huge collection of records into a file structure.
- A B+ tree is a variant of a B-tree that also stores sorted data like a B-tree. The structure of a B-tree is the standard organization for indexes in database systems. A B+ tree is a slightly different data structure that allows sequential processing of data and stores all the data in the lowest level of the tree.

9.7   EXERCISES

9.7.1 Review Questions

Q1. Define:

   a. M-way search tree

   b. B-tree

   c. B+ tree

Q2. Write a difference between B-trees and B+ trees.

Q3. Construct a B-tree of order 3, inserting the keys 10, 20, 50, 60 40, 80, 100, 70, 130, 90, 30, 120, 140, 25, 35, 160, and 180 in a left-to-right sequence. Show the trees after deleting 190 and 60.

Q4. Explain the insertion and deletion of a node in a B-tree.

Q5. Explain B+ tree indexing with the help of an example.

Q6. What do you know about B-trees? Write the steps to create a B-tree. Construct an M-way search tree of order 4 and insert the values 34, 45, 98, 1, 23, 41, 78, 100, 234, 122, 199, 10, and 40.

Q7. Why do we always prefer a higher value of m in a B-tree? Explain.

Q8. Are B-trees of order 2 (i.e., full binary trees)? Explain.

9.7.2 Multiple Choice Questions

Q1. B+ trees are preferred to binary trees in databases because

   a. Disk capacities are greater than memory capacities.

   b. Disk access is slower than memory access.

   c. Disk data transfer rates are less than the memory data transfer rates.

   d. Disks are more reliable than memory.

Q2. In an M-way search tree, M stands for the _____.

    a. Degree of the node   **b.** External nodes

    c. Internal nodes   **d.** None of these

Q3. A B-tree of order 4 is built. What is the maximum number of keys that a node may accommodate before splitting operations take place?

    a. 5   **b.** 2

    c. 4   **d.** 3

Q4. In a B-tree of order m, every node has at the most _____ children.

    a. M + 1   **b.** M − 1

    c. M/2   **d.** M

Q5. What is the best data structure to search the keys in the least amount of time?

    a. B-tree   **b.** M-way search tree

    c. B+ tree   **d.** Binary search tree

Q6. The best case for searching for a value in a binary search tree is

    a. $O(n^2)$ **b.** $O(\log n)$

    c. $O(n)$ **d.** $O(n \log n)$

Q7. External nodes are also called _____.

    a. Index nodes   **b.** Data nodes

    c. Value nodes   **d.** None of the above

Q8. A B+ tree stores redundant keys.

    a. False

    b. True

    c. Not possible to comment

Q9. A B-tree of order 5 can store at least how many keys?

    a. 0   **b.** 1

    c. 2   **d.** 3

HASHING

10.1 INTRODUCTION

In Chapter 6, we discussed three types of search techniques: linear search, binary search, and interpolation search. A linear search has a running time complexity of $O(n)$, whereas a binary search has a running time proportional to $O(\log n)$, where n is equal to the number of elements in the array. The search algorithms discussed in Chapter 6 are efficient. However, their search time is dependent on the number of elements in the array, and none of them can search for an element within the constant time equal to $O(1)$. However, this is very difficult to achieve in search algorithms like the linear search or binary search, as all these algorithms are dependent on the number of elements present in the array. There are many comparisons involved while searching for an element using these types of search algorithms. Therefore, our primary need is to search for the element in a constant time along with few key comparisons. Now, let us take an example. Suppose there is an array of size N and all the keys to be stored in the array are unique and also are in the range 0 to N–1. We store all the records in the array based on the key where the array index and keys are the same. Thus, we can access the records in a constant time with no key comparisons involved. This can be further explained by the following figure:

arr[0]	arr[1]	arr[2]	arr[3]	arr[4]	arr[5]	arr[6]	arr[7]	arr[8]	arr[9]
	1		3	4		6		8	

FIGURE 10.1 An array

In Figure 10.1, there is an array containing five elements. Note that the keys and the array index numbers are the same; that is, the record with the key-value 3 can be directly accessed by array index arr[3]. Similarly, all the records can be accessed through key values and the array index. Thus, this can be done by hashing, where we convert the key into an array index and store the records in the array. This can be done as follows:

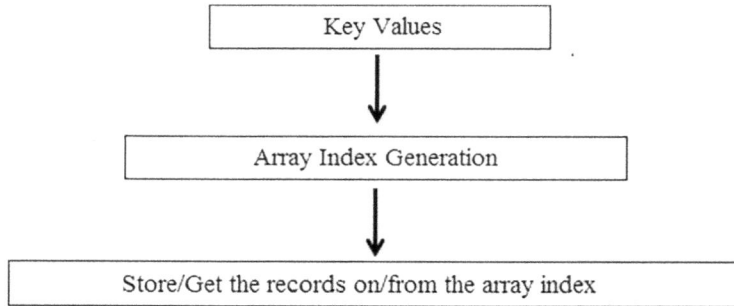

```
┌─────────────────────────────┐
│         Key Values          │
└─────────────────────────────┘
               │
               ▼
┌─────────────────────────────┐
│    Array Index Generation   │
└─────────────────────────────┘
               │
               ▼
┌──────────────────────────────────────────────┐
│  Store/Get the records on/from the array index │
└──────────────────────────────────────────────┘
```

FIGURE 10.2 Array index generation using hashing

The process of array index generation uses a hash function that is used to convert the keys into an array index. The array in which such records are stored is known as a *hash table*.

Practical Application:

1. A simple real-life example is when we search for a word in the dictionary and then find the definition or meaning with the help of a key and its index.

2. Driver's license numbers and insurance card numbers are created using hashing from data items that never change, such as date of birth and name.

Frequently Asked Questions

Q. Explain the term hashing.

Answer:

Hashing is the process of mapping keys to their appropriate locations in the hash table. It is the most effective technique of searching the values in an array or a hash table.

10.1.1 Difference between Hashing and Direct Addressing

In direct addressing, we store the key at the same address as the value of the key, as shown in Figure 10.3. However, in hashing, the address of the key is determined by using a mathematical function known as a hash function, as shown in Figure 10.4. The hash function operates on the key to determine the address of the key. Direct addressing may result in a more random distribution of the key throughout the memory, and hence sometimes leads to a greater waste of space when compared with hashing.

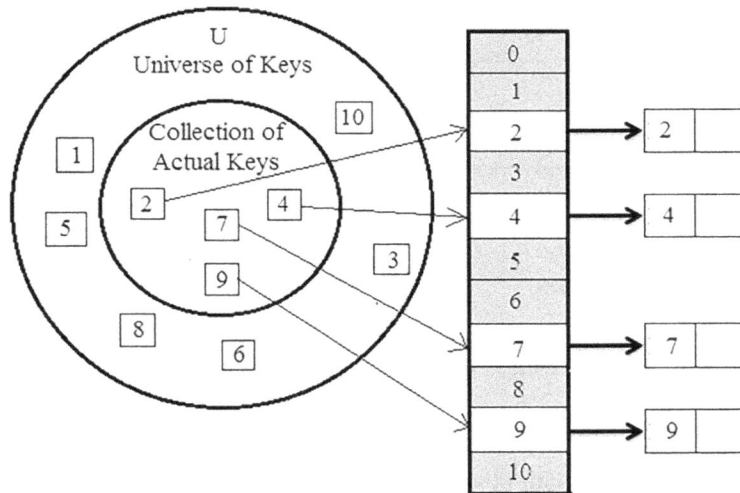

FIGURE 10.3 Mapping of keys using a direct addressing method

10.1.2 Hash Tables

A hash table is a data structure that supports one of the efficient searching techniques, that is, hashing. A hash table is an array in that the data is accessed through a special index called a key. In a hash table, keys are mapped to the array positions by a hash function. A hash function is a function or mathematical formula that, when applied to a key, produces an integer that is used as an index to find a key in the hash table. Thus, a value stored in a hash table can be searched in $O(1)$ time with the help of a hash function. The main idea behind a hash table is to establish the direct mapping between the keys and the indices of the array.

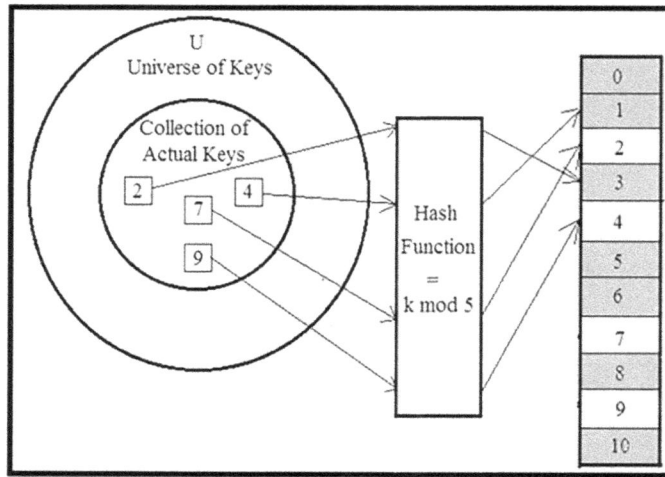

FIGURE 10.4 Mapping of keys to the hash table using hashing

10.1.3 Hash Functions

A hash function is a mathematical formula that, when applied to a key, produces an integer that is used as an index to find a key in the hash table.

Characteristics of the Hash Function

There are four main characteristics of hash functions:

1. The hash function uses all the input data.
2. The hash function must generate different hash values.
3. The hash value is fully determined by the data being hashed.
4. The hash function must distribute the keys uniformly across the entire hash table.

Different Types of Hash Functions

In this section, we discuss some of the common hash functions:

1. **Division Method** – In the division method, a key k is mapped into one of the m slots by taking the remainder of k divided by m. In simple terms, we can say that this method divides an integer, say x, by m, and then uses the remainder so obtained. It is the simplest method of hashing. The hash function is given by

$$h(k) = k \bmod m$$

Address   Key   m no. of slots

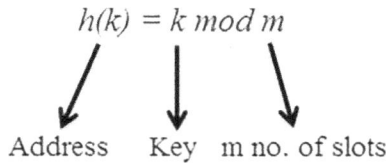

For example, if m = 5 and the key k = 10, then h(k) = 2. Thus, the division method works very fast, as it requires only a single division operation. Although this method is good for any value of m, consider that if m is an even number then h(k) is even when the value of k is even, and similarly h(k) is odd when the value of k is odd. Therefore, if the even and odd keys are almost equal, then there will be no problem. But if there is a larger number of even keys, then the division method is not good, as it will not distribute the keys uniformly in the hash table. We also avoid certain values of m; that is, m should not be a power of 2, because if h(k) = k mod 2^x, then h(k) will extract the lowest x bits of k. The main drawback of the division method is that many consecutive keys map to consecutive hash values, which means that consecutive array locations are occupied, and hence there is an effect on the performance.

Frequently Asked Questions

Q. Given a hash table of 50 memory locations, calculate the hash values of keys 20 and 75 using the division method.

Answer:

m = 50, k1 = 10, k2 = 75 hash values are calculated as

h(10) = 10 % 50 = 10

h(75) = 75 % 50 = 25

---

2. **Mid Square Method** – In the mid square method, we calculate the square of the given key. After getting the number, we extract some digits from the middle of that number as an address.

   For example, if key k = 5025, then k^2 = 25250625. Thus, h(5025) = 50.

   This method works very well, as all the digits of the key contribute to the output; that is, all the digits contribute to producing the middle digits. The same r digits must be chosen from all the keys in this method.

Frequently Asked Questions

Q. Given a hash table of 100 memory locations, calculate the hash values of keys 2045 and 1357 using the mid square method.

Answer:

There are 100 memory locations where indices are from 0 to 99. Hence, only two digits are taken to map the keys. So, the value of r is equal to 2.

$k_1 = 2045, k^2 = 4182025, h(2045) = 20$

$k_2 = 1357, k^2 = 1841449, h(1357) = 14$

Note: The third and fourth digits are chosen to start from the right.

3. **Folding Method** – In the folding method, we break the key into pieces such that each piece has the same number of digits except the last one, which may have fewer digits as compared to the other pieces. Now, these individual pieces are added. We ignore the carry if it exists. Hence, the hash value is formed.

For example, if m = 100 and the key k = 12345678, then the indices will vary from 0 to 99, and thus each piece of the key must have two digits. Therefore, the given key is broken into four pieces, that is, 12, 34, 56, and 78. We add all these, that is, 12 + 34 + 56 + 78 = 180. Thus, the hash value will be 80 (ignore the last carry).

Frequently Asked Questions

Q. Given a hash table of 100 memory locations, calculate the hash values of keys 2486 and 179 using the folding method.

Answer:

There are 100 memory locations where indices are from 0 to 99. Hence, each piece of the key must have two digits.

$h(2486) = 24 + 86 = 110$

$h(2486) = 10$ *(ignore the last carry)*

$h(179) = 17 + 9 = 26$

$h(179) = 26$

10.1.4 Collision

A collision is a situation that occurs when a hash function maps two different keys to a single/same location in the hash table. Suppose we want to store a record at one location. Another record cannot be stored at the same location as it is obvious that two records cannot be stored at the same location. Thus, there are methods to solve this problem, which are called collision resolution techniques.

10.1.5 Collision Resolution Techniques

Collision resolution techniques are used to overcome the problem of collision in hashing. Two popular methods are used for resolving collisions:

1. Collision Resolution by Chaining Method

2. Collision Resolution by Open Addressing Method

10.1.5.1 Chaining Method

In the chaining method, a chain of elements is maintained where all the elements have the same hash address. The hash table here behaves like an array of references. Each location in the hash table stores a reference to the linked list, which contains all the key elements that were hashed to that location. For example, location 5 in the hash table points to the key values that hashed to location 5. If no key-value hashes to location 5, then location 5 contains NULL. Figure 10.5 shows how the key values are mapped to the hash table and how they are stored in the linked list.

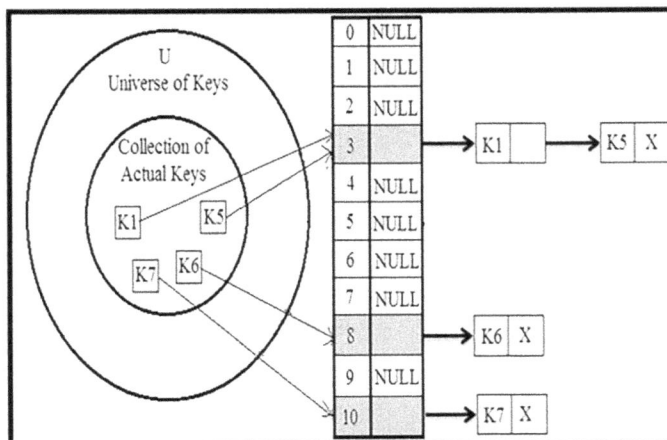

FIGURE 10.5 Keys hashed by the chaining method

Operations on a Chained Hash Table

1. **Insertion in a Chained Hash Table** – The process of inserting an element is quite simple. First, we get the hash value from the hash function that maps to the hash table. After mapping, the element is inserted in the linked list. The running time complexity of inserting an element in a chained hash table is O(1).

2. **Deletion from a Chained Hash Table** – The process of deleting an element from the chained hash table is the same as we used in the singly linked list. First, we perform a search operation, and then the delete operation as in the case of the singly linked list is performed. The running time complexity of deleting an element from a chained hash table is O(m), where m is the number of elements present in the linked list at that location.

3. **Searching in a Chained Hash Table** – The process of searching for an element in a chained hash table is also very simple. First, we get the hash value of the key by the hash function in the hash table. Then we search for the element in the linked list. The running time complexity of searching for an element in a chained hash table is O(m), where m is the number of elements present in the linked list at that location.

Frequently Asked Questions

Q. Insert the keys 4, 9, 20, 35, and 49 in a chained hash table of 10 memory locations. Use hash function h(k) = k mod m.

Answer:

Initially, the hash table is given as

0	NULL
1	NULL
2	NULL
3	NULL
4	NULL
5	NULL
6	NULL
7	NULL
8	NULL
9	NULL

Now, we insert 4 in the hash table.

Step 1:

Key to be inserted = 4

h(4) = 4 mod 10

h(4) = 4

We create a linked list for location 4, and the key element 4 is stored in it.

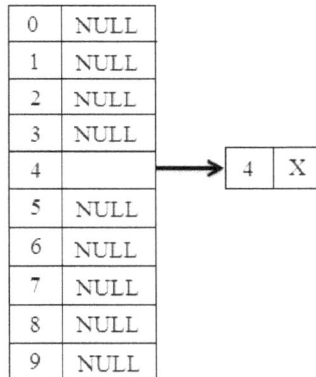

0	NULL
1	NULL
2	NULL
3	NULL
4	→ 4 | X
5	NULL
6	NULL
7	NULL
8	NULL
9	NULL

Step 2:

Key to be inserted = 9

h(9) = 9 mod 10

h(9) = 9

We create a linked list for location 9, and the key element 9 is stored in it.

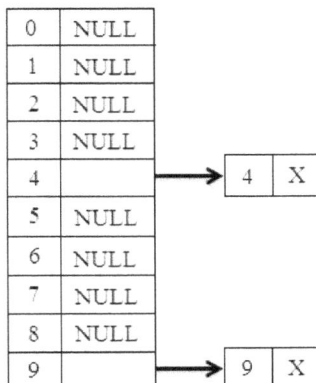

0	NULL
1	NULL
2	NULL
3	NULL
4	→ 4 | X
5	NULL
6	NULL
7	NULL
8	NULL
9	→ 9 | X

Step 3:

Key to be inserted = 20

h(20) = 20 mod 10

h(20) = 2

We create a linked list for location 2, and the key element 20 is stored in it.

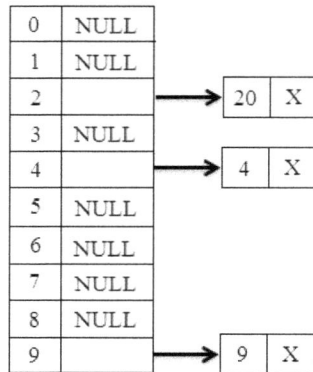

0	NULL
1	NULL
2	→ 20 | X
3	NULL
4	→ 4 | X
5	NULL
6	NULL
7	NULL
8	NULL
9	→ 9 | X

Step 4:

Key to be inserted = 35

h(35) = 35 mod 10

h(35) = 5

We create a linked list for location 5, and the key element 35 is stored in it.

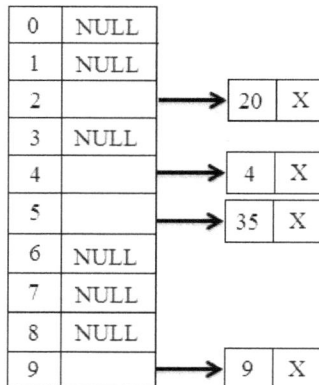

0	NULL
1	NULL
2	→ 20 | X
3	NULL
4	→ 4 | X
5	→ 35 | X
6	NULL
7	NULL
8	NULL
9	→ 9 | X

Step 5:

Key to be inserted = 49

h(49) = 49 mod 10

h(49) = 9

We insert 49 at the end of the linked list of location 9.

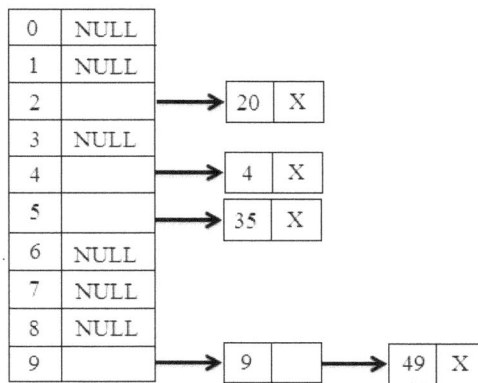

Advantages and Disadvantages of the Chained Method

The main advantage of this method is that it completely resolves the problem of collision. It remains effective even when the key elements to be stored in the hash table are higher than the number of locations in the hash table. However, it is quite obvious that with the increase in the number of key elements, the performance of this method decreases.

The disadvantage of this method is the waste of storage space, as the key elements are stored in the linked list; also, the references are required for each element to get accessed, which in turn consumes more space.

10.1.5.2 Open Addressing Method

In the open addressing method, all the elements are stored in the hash table itself. Once a collision takes place, open addressing computes new locations using the probe sequence, and the next element or next record is stored on that location. Probing is the process of examining the memory locations in the hash table. When we perform the insertion operation in the open addressing method, we first successively probe/examine the hash table until we find an empty slot in that the new key can be inserted. The open addressing method can be implemented using

- Linear Probing
- Quadratic Probing
- Double Hashing

Linear Probing

Linear probing is the simplest approach to resolving the problem of collision in hashing. In this method, if a key is already stored at a location generated by the hash function h(k), then the situation can be resolved by the following hash function:

$$h'(k) = (h(k) + i) \bmod m$$

$$\text{where } h(k) = k \bmod m$$

$$i = \text{Probe no.} = 0, 1, 2, 3..........(m\text{ -}1)$$

$$m = \text{no. of slots}$$

Now, let us examine how this technique works. For a given key k, first, the location generated by (h(k) + 0) mod m is probed, because for the first time i = 0. If the location generated is free, then the key is stored in it. Otherwise, the second probe is generated for i = 1 given by the hash function (h(k) + 1) mod m. Similarly, if the location generated is free, then the key is stored in it; otherwise, subsequent probes are generated such as (h(k) + 2) mod m, (h(k) + 3) mod m, and so on, until we find a free location.

Frequently Asked Questions

Q. Given keys k = 13, 25, 14, and 35 maps these keys into a hash table of size m = 5 using linear probing.

Answer:

Initially, the hash table is given as

0	1	2	3	4
NULL	NULL	NULL	NULL	NULL

Step 1:

i = 0

Key to be inserted = 13

h'(k) = (k mod m + i) mod m

h'(13) = (13 % 5 + 0) % 5

h'(13) = (3 + 0) % 5

h'(13) = 3 % 5 = 3

Since location T[3] is free, 13 is inserted at location T[3].

0	1	2	3	4
NULL	NULL	NULL	13	NULL

Step 2:

i = 0

Key to be inserted = 25

h'(25) = (25 % 5 + 0) % 5

h'(25) = (0 + 0) % 5

h'(13) = 0 % 5 = 0

Since location T[0] is free, 25 is inserted at location T[0].

0	1	2	3	4
25	NULL	NULL	13	NULL

Step 3:

i = 0

Key to be inserted = 14

h'(14) = (14 % 5 + 0) % 5

h'(14) = (4 + 0) % 5

$h'(14) = 4 \% 5 = 4$

Since location T[4] is free, 14 is inserted at location T[4].

0	1	2	3	4
25	NULL	NULL	13	14

Step 4:

$i = 0$

Key to be inserted = 35

$h'(35) = (35 \% 5 + 0) \% 5$

$h'(35) = (0 + 0) \% 5$

$h'(35) = 0 \% 5 = 0$

Since location T[0] is not free, the next probe sequence, that is, i = 1, is computed as

$i = 1$

$h'(35) = (35 \% 5 + 1) \% 5$

$h'(35) = (0 + 1) \% 5$

$h'(35) = 1 \% 5 = 1$

Since location T[1] is free, 35 is inserted at location T[1].

Thus, the final hash table is

0	1	2	3	4
25	35	NULL	13	14

Here is a program to show the linear probing technique of the collision resolution method.

```
# Program to implement hashing with linear probing

class hashTable:
    # initialize hash Table
    def __init__(self):
        self.size = int(input("Enter the size of the hash table : "))
        # initialize table with all elements 0
        self.table = list(0 for i in range(self.size))
        self.elementCount = 0
        self.comparisons = 0

    # method that checks if the hash table is full or not
    def isFull(self):
        if self.elementCount == self.size:
            return True
        else:
            return False

    # method that returns position for a given element
    def hashFunction(self, element):
        return element % self.size

    # method that inserts element inside the hash table
    def insert(self, element):
        # checking if the table is full
        if self.isFull():
            print("Hash Table Full")
            return False

        isStored = False

        position = self.hashFunction(element)

        # checking if the position is empty
        if self.table[position] == 0:
            self.table[position] = element
            print("Element " + str(element) + " at position " + str(position))
            isStored = True
            self.elementCount += 1

        # collision occurred hence we do linear probing
        else:
            print("Collision has occurred for element " +
str(element) + " at position " + str(position) + " finding
new position.")
            while self.table[position] != 0:
                position += 1
                if position >= self.size:
                    position = 0
```

```python
                self.table[position] = element
                isStored = True
                self.elementCount += 1
        return isStored
    # method that searches for an element in the table
    # returns position of element if found
    # else returns False
    def search(self, element):
        found = False

        position = self.hashFunction(element)
        self.comparisons += 1

        if(self.table[position] == element):
            return position
            isFound = True

        # if element is not found at position returned hash function
        # then first we search element from position+1 to end
        # if not found then we search element from position-1 to 0
        else:
            temp = position - 1
          # check if the element is stored between position+1 to size
            while position < self.size:
                if self.table[position] != element:
                    position += 1
                    self.comparisons += 1
                else:
                    return position

          # now checking if the element is stored between position-1 to 0
            position = temp
            while position >= 0:
                if self.table[position] != element:
                    position -= 1
                    self.comparisons += 1
                else:
                    return position

        if not found:
            print("Element not found")
            return False
    # method to remove an element from the table
    def remove(self, element):
        position = self.search(element)
        if position is not False:
            self.table[position] = 0
```

```
            print("Element " + str(element) + " is deleted")
            self.elementCount -= 1
        else:
            print("Element is not present in the hash table")
        return

    # method to display the hash table
    def display(self):
        print("\n")
        for i in range(self.size):
            print(str(i) + " = " + str(self.table[i]))
        print("The number of elements in the table are : " +
str(self.elementCount))
```

The output of the program is as follows:

```
>>> hash1=hashTable()
Enter the size of the hash table : 6
>>> hash1.insert(10)
Element 10 at position 4
True
>>> hash1.insert(21)
Element 21 at position 3
True
>>> hash1.insert(4)
Collision has occurred for element 4 at position 4 finding new position.
True
>>> hash1.insert(3)
Collision has occurred for element 3 at position 3 finding new position.
True
```

```
>>> hash1.display()


0 = 3
1 = 0
2 = 0
3 = 21
4 = 10
5 = 4
The number of elements in the table are : 4
>>> hash1.search(3)
0
>>> hash1.remove(21)
Element 21 is deleted
```

```
>>> hash1.display()


0 = 3
1 = 0
2 = 0
3 = 0
4 = 10
5 = 4
The number of elements in the table are : 3
```

Advantages and Disadvantages of Linear Probing

Linear probing is a very useful technique, as the algorithm provides reliable memory caching through a good locality of the address. However, the main disadvantage of this method is that it results in clustering. Due to clustering, there is a higher risk of collisions taking place. The time required for searching also increases with the size of the clusters. We can say that the higher the number of collisions requires a higher number of probes to find a vacant location, and the performance is decreased. This is known as *primary clustering*. We can avoid this clustering by using other techniques, such as quadratic probing and double hashing.

Quadratic Probing

Quadratic probing is another approach to resolving the problem of collision in hashing. In this method, if a key is already stored at a location generated by the hash function h(k), then the situation can be resolved by the following hash function:

$$h'(k) = (h(k) + c_1 i + c_2 i^2) \mod m$$

$$\text{where } h(k) = k \mod m$$

$$i = \text{Probe no.} = 0, 1, 2, 3.........(m-1)$$

$$c_1, c_2 = consonants$$

$$(c_1, c_2 \text{ should not be equal to zero})$$

The quadratic probing method is better than linear probing, as it terminates the phenomenon of primary clustering because of its searching speed; that is, it is doing a quadratic search. For a given key k, first the location generated by (h(k) + 0 + 0) mod m is probed, because for the first time i = 0. If the location generated is free, then the key is stored in it. Otherwise, subsequent positions probed are offset by the amounts/factors that depend in a quadratic manner on the probe number i. The quadratic probing method works better than linear probing, but to maximize the use of the hash table, the values of m, c_1, and c_2 are constrained.

Frequently Asked Questions

Q. Given keys k = 25, 13, 14, and 35, map these keys into a hash table of size m = 5 using quadratic probing with c_1 = 1 and c_2 = 3.

Answer:

Initially, the hash table is given as follows:

0	1	2	3	4
NULL	NULL	NULL	NULL	NULL

Step 1:

i = 0

$c_1 = 1, c_2 = 3$

Key to be inserted = 25

$h'(k) = (k \bmod m + c_1 i + c_2 i^2) \bmod m$

$h'(25) = (25 \% 5 + 1 \times 0 + 3 \times (0)^2) \% 5$

$h'(25) = (0 + 0) \% 5$

$h'(13) = 0 \% 5 = 0$

Since location T[0] is free, 25 is inserted at location T[0].

0	1	2	3	4
25	NULL	NULL	NULL	NULL

Step 2:

i = 0

$c_1 = 1, c_2 = 3$

Key to be inserted = 13

$h'(13) = (13 \% 5 + 1 \times 0 + 3 \times (0)^2) \% 5$

$h'(13) = (3 + 0) \% 5$

$h'(13) = 3 \% 5 = 3$

Since location T[3] is free, 13 is inserted at location T[3].

0	1	2	3	4
25	NULL	NULL	13	NULL

Step 3:

$i = 0$

$c_1 = 1, c_2 = 3$

Key to be inserted = 14

$h'(14) = (14 \% 5 + 1 \times 0 + 3 \times (0)^2) \% 5$

$h'(14) = (4 + 0) \% 5$

$h'(14) = 4 \% 5 = 4$

Since location T[4] is free, 14 is inserted at location T[4].

0	1	2	3	4
25	NULL	NULL	13	14

Step 4:

$i = 0$

$c_1 = 1, c_2 = 3$

Key to be inserted = 35

$h'(35) = (35 \% 5 + 1 \times 0 + 3 \times (0)^2) \% 5$

$h'(35) = (0 + 0) \% 5$

$h'(35) = 0 \% 5 = 0$

Since location T[0] is not free, the next probe sequence, that is, $i = 1$, is computed as

$i = 1$

$h'(35) = (35 \% 5 + 1 \times 1 + 3 \times (1)^2) \% 5$

$h'(35) = (0 + 1 + 3) \% 5$

$h'(35) = 4 \% 5 = 4$

Again, since location T[4] is not free, the next probe sequence, that is, i = 2, is computed as

$i = 2$

$h'(35) = (35 \% 5 + 1 \times 2 + 3 \times (2)^2) \% 5$

$h'(35) = (0 + 2 + 12) \% 5$

$h'(35) = 14 \% 5 = 4$

Again, since location T[4] is not free, the next probe sequence, that is, i = 3, is computed as

$i = 3$

$h'(35) = (35 \% 5 + 1 \times 3 + 3 \times (3)^2) \% 5$

$h'(35) = (0 + 3 + 27) \% 5$

$h'(35) = 30 \% 5 = 0$

Again, since location T[0] is not free, the next probe sequence, that is, i = 4, is computed as

$i = 4$

$h'(35) = (35 \% 5 + 1 \times 4 + 3 \times (4)^2) \% 5$

$h'(35) = (0 + 4 + 48) \% 5$

$h'(35) = 52 \% 5 = 2$

Since location T[2] is free, 35 is inserted at location T[2].

Thus, the final hash table is as follows:

0	1	2	3	4
25	NULL	35	13	14

Here is a program to show the quadratic probing technique of the collision resolution method.

```python
# Program to implement hashing with quadratic probing

class hashTable:
    # initialize hash Table
    def __init__(self):
        self.size = int(input("Enter the size of the hash table : "))
        # initialize table with all elements 0
        self.table = list(0 for i in range(self.size))
        self.elementCount = 0
        self.comparisons = 0

    # method that checks if the hash table is full or not
    def isFull(self):
        if self.elementCount == self.size:
            return True
        else:
            return False

    # method that returns position for a given element
    # replace with your own hash function
    def hashFunction(self, element):
        return element % self.size

    # method to resolve collision by quadratic probing method
    def quadraticProbing(self, element, position):
        posFound = False
        # limit variable is used to restrict the function
        # from going into infinite loop
        # limit is useful when the table is 80% full
        limit = 50
        i = 1
        # start a loop to find the position
        while i <= limit:
            # calculate new position by quadratic probing
            newPosition = position + (i**2)
            newPosition = newPosition % self.size
            # if newPosition is empty then break out of loop
            # and return new Position
            if self.table[newPosition] == 0:
                posFound = True
                break
            else:
                # as the position is not empty increase i
                i += 1
        return posFound, newPosition

    # method that inserts element inside the hash table
    def insert(self, element):
        # checking if the table is full
        if self.isFull():
            print("Hash Table Full")
```

```
            return False

        isStored = False

        position = self.hashFunction(element)

        # checking if the position is empty
        if self.table[position] == 0:
            # empty position found , store the element and
print the message
            self.table[position] = element
            print("Element " + str(element) + " at position "
     + str(position))
            isStored = True
            self.elementCount += 1

        # collision occurred hence we do linear probing
        else:
            print("Collision has occurred for element " + str(element)
+ " at position " + str(position) + " finding new Position.")
            isStored, position = self.quadraticProbing(element, position)
            if isStored:
                self.table[position] = element
                self.elementCount += 1

        return isStored

    # method that searches for an element in the table
    # returns position of element if found
    # else returns False
    def search(self, element):
        found = False

        position = self.hashFunction(element)
        self.comparisons += 1
        if(self.table[position] == element):
            return position

        # if element is not found at position returned hash function
        # then we search element using quadratic probing
        else:
            limit = 50
            i = 1
            newPosition = position
            # start a loop to find the position
            while i <= limit:
                # calculate new position by quadratic probing
                newPosition = position + (i**2)
                newPosition = newPosition % self.size
                self.comparisons += 1

                # if element at newPosition is equal to the
required element
```

```
                if self.table[newPosition] == element:
                    found = True
                    break

                elif self.table[newPosition] == 0:
                    found = False
                    break

                else:
                    # as the position is not empty increase i
                    i += 1
            if found:
                return newPosition
            else:
                print("Element not found")
                return found

    # method to remove an element from the table
    def remove(self, element):
        position = self.search(element)
        if position is not False:
            self.table[position] = 0
            print("Element " + str(element) + " is deleted")
            self.elementCount -= 1
        else:
            print("Element is not present in the hash table")
        return

    # method to display the hash table
    def display(self):
print("\n")
        for i in range(self.size):
            print(str(i) + " = " + str(self.table[i]))
        print("The number of elements in the table are : " +
str(self.elementCount))
```

The output of the program is as follows:

```
>>> hash2=hashTable()
Enter the size of the hash table : 6
>>> hash2.insert(10)
Element 10 at position 4
True
>>> hash2.insert(5)
Element 5 at position 5
True
>>> hash2.insert(4)
Collision has occurred for element 4 at position 4 finding new Position.
True
>>> hash2.insert(23)
Collision has occurred for element 23 at position 5 finding new Position.
True
```

```
>>> hash2.display()

0 = 23
1 = 0
2 = 4
3 = 0
4 = 10
5 = 5
The number of elements in the table are : 4
>>> hash2.search(4)
2
>>> hash2.remove(10)
Element 10 is deleted
```

```
>>> hash2.display()

0 = 23
1 = 0
2 = 4
3 = 0
4 = 0
5 = 5
The number of elements in the table are : 3
```

Advantages and Disadvantages of Quadratic Probing

As previously discussed, one of the biggest advantages of quadratic probing is that it eliminates the phenomenon of primary clustering. Yet one of the major disadvantages of this method is that a sequence of successive probes may only cover some portion of the hash table, and this portion may be quite small. Therefore, if such a situation occurs, then it will be difficult for us to find an empty location in the hash table, even though the table is not full. Hence, quadratic probing encounters a problem that is known as *secondary clustering*. In this method, the chance of multiple collisions increases as the hash table becomes full. This type of situation can be overcome by double hashing.

Double Hashing

Double hashing is one of the best methods available for open addressing. As the name suggests, this method uses two hash functions to operate rather than a single hash function. The hash function is given as follows:

$$h'(k) = (h_1(k) + ih_2(k)) \ mod \ m,$$

where $h_1(k) = k \bmod m$ and $h_2(k) = k \bmod m'$ are the two hash functions, m is the size of the hash table, m' is less than m (can be $(m-1)$ or $(m-2)$), and i is the probe number that varies from 0 to $(m-1)$.

Now, let's consider how this technique works. For a given key k, first, the location generated by $(h_1(k) \bmod m)$ is probed, because for the first time i = 0. If the location generated is free, then the key is stored in it. Otherwise, subsequent probes generate locations that are at an offset of $(h_2(k) \bmod m)$ from the previous location. The offset may vary with every probe depending on the value generated by the second hash function, that is, $(h_2(k) \bmod m)$. As a result, the performance of double hashing is very near to the performance of the "ideal" scheme of uniform hashing.

Frequently Asked Questions

Q. Given keys k = 71, 29, 38, 61, and 100, map these keys into a hash table of size m = 5 using double hashing. Use h_1 = (k mod 5) and h_2 = (k mod 4).

Answer:

Initially, the hash table is given as

0	1	2	3	4
NULL	NULL	NULL	NULL	NULL

Step 1:

i = 0

Key to be inserted = 71

$h'(k) = (h_1(k) + ih_2(k)) \bmod m$

$h'(k) = (k \bmod m + (i\,k \bmod m')) \bmod m$

$h'(71) = (71 \% 5 + (0 \times 71 \% 4)) \% 5$

$h'(71) = (1 + (0 \times 3)) \% 5$

$h'(71) = 1 \% 5 = 1$

Since location T[1] is free, 71 is inserted at location T[1].

0	1	2	3	4
NULL	71	NULL	NULL	NULL

Step 2:

i = 0

Key to be inserted = 29

h'(k) = (k mod m + (i k mod m')) mod m

h'(29) = (29 % 5 + (0 X 29 % 4)) % 5

h'(29) = (4 + (0 X 1)) % 5

h'(29) = 4 % 5 = 4

Since location T[4] is free, 29 is inserted at location T[4].

0	1	2	3	4
NULL	71	NULL	NULL	29

Step 3:

i = 0

Key to be inserted = 38

h'(k) = (k mod m + (i k mod m')) mod m

h'(38) = (38 % 5 + (0 X 38 % 4)) % 5

h'(38) = (3 + (0 X 2)) % 5

h'(38) = 3 % 5 = 3

Since location T[3] is free, 38 is inserted at location T[3].

0	1	2	3	4
NULL	71	NULL	38	29

Step 4:

i = 0

Key to be inserted = 61

h'(k) = (k mod m + (i k mod m')) mod m

h'(61) = (61 % 5 + (0 X 61 % 4)) % 5

h'(61) = (1 + (0 X 1)) % 5

h'(61) = 1 % 5 = 1

Since location T[1] is not free, the next probe sequence, that is, i = 1, is computed as

$i = 1$

$h'(61) = (61 \% 5 + (1 \times 61 \% 4)) \% 5$

$h'(61) = (1 + (1 \times 1)) \% 5$

$h'(61) = (1 + 1) \% 5$

$h'(61) = 2 \% 5 = 2$

Since location T[2] is free, 61 is inserted at location T[2].

0	1	2	3	4
NULL	71	61	38	29

Step 5:

$i = 0$

Key to be inserted = 100

$h'(k) = (k \bmod m + (i \, k \bmod m')) \bmod m$

$h'(100) = (100 \% 5 + (0 \times 100 \% 4)) \% 5$

$h'(100) = (0 + (0 \times 0)) \% 5$

$h'(100) = 0 \% 5 = 0$

Since location T[0] is free, 100 is inserted at location T[0].

Thus, the final hash table is as follows:

0	1	2	3	4
100	71	61	38	29

Advantages and Disadvantages of Double Hashing

The double hashing method is free from all the problems of primary clustering and secondary clustering. It also minimizes repeated collisions.

10.2 SUMMARY

- A hash table is an array in that the data is accessed through a special index called a key. In a hash table, keys are mapped to the array positions by a hash function.

- A hash function is a mathematical formula that, when applied to a key, produces an integer that is used as an index to find a key in the hash table.

- There are different types of hash functions that use numeric keys. Popular methods are the division method, the mid square method, and the folding method.

- In the division method, a key k is mapped into one of the m slots by taking the remainder of k divided by m. The main drawback of the division method is that many consecutive keys map to consecutive hash values respectively, that means that consecutive array locations will be occupied, and hence there will be an effect on the performance.

- In the mid square method, we calculate the square of the given key. After getting the number, we extract some digits from the middle of that number as an address.

- In the folding method, we break the key into pieces such that each piece has the same number of digits except the last one, which may have lower digits as compared to other pieces. Now, these individual pieces are added. Hence, the hash value is formed.

- A collision is a situation that occurs when a hash function maps two different keys to a single/same location in the hash table.

- Collision resolution techniques are used to overcome the problem of collision in hashing. Two popular methods are used for resolving collisions, which are collision resolution by the chaining method and collision resolution by the open addressing method.

- In the chaining method, a chain of elements is maintained where the elements have the same hash address. Each location in the hash table stores an address to the linked list that contains all the key elements that were hashed to that location. The disadvantage of this method is the waste of storage space, as the key elements are stored in the linked list; addresses are required for each element to get accessed, which in turn consumes more space.

- In an open addressing method, all the elements are stored in the hash table itself. There is no need to provide the address in this method, that is the biggest advantage of this method. Once a collision takes place, open addressing computes new locations using the probe sequence, and the next element or next record is stored in that location.

- Probing is the process of examining the memory locations in the hash table.

- Linear probing is the simplest approach to resolving the problem of collision in hashing. In this method, if a key is already stored at a location generated by the hash function h(k), then the situation can be resolved by the following hash function:

$$h'(k) = (h(k) + i) \bmod m$$

- Quadratic probing is another approach to resolving the problem of collision in hashing. In this method, if a key is already stored at a location generated by the hash function h(k), then the situation can be resolved by the following hash function:

$$h'(k) = (h(k) + c_1 i + c_2 i^2) \bmod m$$

- Double hashing is one of the best methods available for open addressing. As the name suggests, this method uses two hash functions to operate rather than a single hash function. The hash function is

$$h'(k) = (h_1(k) + i h_2(k)) \bmod m$$

10.3 EXERCISES

10.3.1 Review Questions

Q1. What are hash tables?

Q2. What is hashing? Give some of its practical applications.

Q3. Define the hash function and also explain the various characteristics of a hash function.

Q4. What is a collision in hashing and how it can be resolved?

Q5. Explain the different types of hash functions along with examples.

Q6. Discuss the collision resolution techniques in hashing.

Q7. What is clustering in hashing? What are the two types of clustering?

Q8. What do you understand about double hashing?

Q9. Define the following terms:

  a. Quadratic Probing

  b. Linear Probing

Q10. What is the chaining method in hashing and how it can help in resolving collisions?

Q11. Consider a hash table of size 10. Using linear probing, insert the keys 12, 45, 67, 122, 78, and 34 into it.

Q12. Consider a hash table of size 9. Using double hashing, insert the keys 4, 17, 30, 55, 90, 11, 54, and 77 into it. Take h_1 = k mod 9 and h_2 = k mod 6.

Q13. Consider a hash table of size 11. Using quadratic probing, insert the keys 10, 45, 56, 97, 123, and 1 into it.

Q14. How can the open addressing method be used in resolving collisions?

Q15. Write a Python function to retrieve an item from the hash table using linear probing and quadratic probing.

10.3.2 Multiple Choice Questions

Q1. Which of the following collision resolution techniques is free from the clustering phenomenon?

  a. Linear Probing

  b. Quadratic Probing

  c. Double Hashing

  d. None of these

Q2. The process of examining a memory location is called _____.

  a. Probing

  b. Hashing

  c. Chaining

  d. Addressing

Q3. A hash table with chaining as a collision resolution technique degenerates to a

   a. Tree   **b.** Graph

   c. Array   **d.** Linked List

Q4. Which of the probing techniques suffers from the problem of primary clustering?

   a. Quadratic Probing   **b.** Linear Probing

   c. Double Hashing   **d.** All of these

Q5. Given the hash function $h(k) = k \bmod 6$, what is the number of collisions to store the following sequence of keys, 16, 20, 45, 68 using open addressing?

   a. 1   **b.** 3

   c. 2   **d.** 5

Q6. In a hash table, an element with the key k is stored at _____.

   a. k   **b.** $h(k^2)$

   c. $h(k)$ **d.** $\log h(k)$

Q7. A good hash function eliminates the problem of collision.

   a. True

   b. False

   c. Not possible to comment

Q8. Given the hash function of size 7 and hash function $h(k) = k \bmod 7$, what is the number of collisions with linear probing for the insertion of the following keys: 29, 36, 16, and 30?

   a. 1   **b.** 2

   c. 3   **d.** 4

Q9. _____ is the process of mapping keys to appropriate locations in the hash table.

   a. Probing   **b.** Hashing

   c. Collision   **d.** Addressing

FILES

11.1 INTRODUCTION

In most organizations, a large amount of data is collected in one form or another. Some organizations use various types of data collection applications for collecting the data. When we talk about an organization, it is not only the large ones like schools, colleges, and companies, but also small companies, like the bakery on the corner. The collection and exchange of data takes place everywhere. For example, when we get admitted into a school, a lot of data is collected by the school, such as name, age, address, parent's name, and blood type. We know that in the past, data was collected in the form of paper documents, which were very difficult to handle and store. Therefore, to efficiently and effectively analyze the collected data, computers are used to store the data in the form of files. A file, in computer terminology, is defined as a block of useful data in a persistent storage medium; that is, the file is available for future use. Data is organized in a hierarchical order in files. The hierarchical order includes items such as records, fields, and so forth, which all are defined as follows.

11.2 TERMINOLOGY

- **Data field** – A *data field* is a unit which stores a unary fact. It is usually characterized by its type and size. For example, "employee's name" is a data field that stores the names of employees.
- **Record** – The collection of related data fields is called a *record*. For example, an employee's record may contain various data fields such as name, ID, address, and contact number.

- **File** – The collection of related records is called a *file*. An example is a file of the employees working in an organization.
- **Directory** – The collection of related files is called a *directory*. Every file in a computer system is stored in a directory.
- **File Name** – The name of a file is a string of characters.
- **Read-only** – A file named *read-only* cannot be modified or deleted. If we try to delete the file, then a particular message is displayed.
- **Hidden** – A file marked as *hidden* is not displayed in the directory.

11.3 FILE OPERATIONS

There are various operations that can be performed on files.

1. **File Creation** – This is the first operation to be performed on the files if the file has not been created. A file is started by specifying its name and mode. The records are inserted into the file by opening the file in writing mode. Once all the records are inserted into the file, the file can be used for future read and write operations. For example, we create a new file named EMPLOYEE.

2. **Updating a File** – This means changing the contents of a file. It is usually done in the following ways:

   a. **Inserting into a File** – The new record is placed into the file. For example, if a new employee joins an organization, his/her record is inserted in the EMPLOYEE file.

   b. **Modifying a File** – The existing records are altered in the file. For example, if the address of an employee is changed, then the new address must be modified in the EMPLOYEE file.

   c. **Deleting from a File** – The existing record is removed from the file. For example, if an employee quits a job, then his/her record is deleted from the EMPLOYEE file.

3. **Retrieving from a File** – This refers to the process of extracting some useful data from a file. It is usually done in the following ways:

   a. **Enquiring** – This retrieves a small amount of data from a file.

   b. **Generating a Report** – This retrieves a huge amount of data from a file.

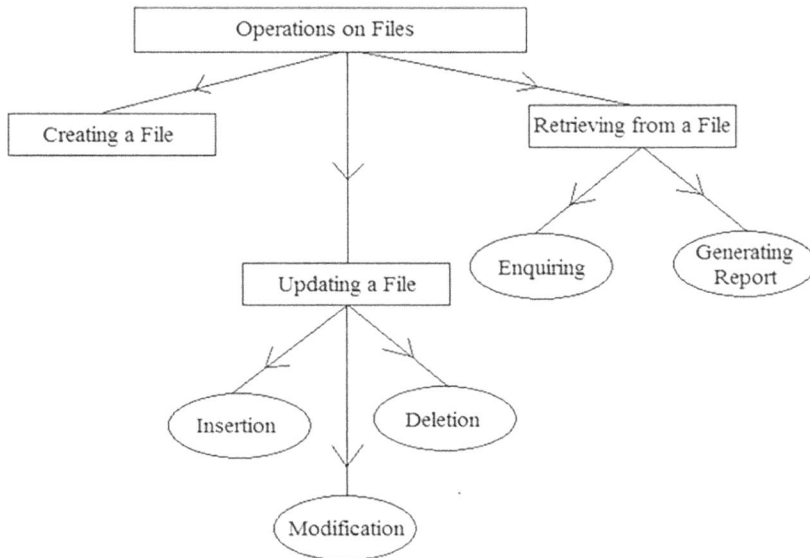

FIGURE 11.1 Operations on files

11.4 FILE CLASSIFICATION

A file is classified into two types:

1. **Text Files** – A *text file*, often called a flat-file, is a file that stores all the numeric or non-numeric data using its corresponding ASCII values. The data can be a string of letters, numbers, or special symbols. Therefore, it is also known as an ASCII file. Usually, a text file has a special marker known as the end of file marker which denotes the end of the file.

2. **Binary Files** – A *binary file* is a file that contains all the data in the binary form of 1s and 0s. It stores the data in the same form as that of primary memory. Thus, a binary file is not readable by human beings. Binary files are read by computer programs, and they decode the binary files into something meaningful. Data is efficiently stored in binary files.

11.5 C vs. C++ vs. JAVA vs. PYTHON FILE HANDLING

File handling is an important process, and one must be aware of the file handling process irrespective of any language. This is especially true when it comes to C vs. C++vs. Java vs. Python File Handling, because it becomes difficult to understand the operations and processes on files as these languages possess similar kinds of functions/operators. Hence, there are some points to remember while working with files, which are shown in the table.

Table 11.1 Comparison of how four programming languages approach files

C	C++	Java	Python
In C, fopen, fclose, fwrite, fread, fseek, fprint, fscanf, and various other functions are called directly without any help of an object.	In C++, open, close, and other functions are called with the help of an object that is, for example, fstream f. Here, f is the object of the stream class, and all the functions are called with the help of objects like f.open, f.close, f.read, and f.write.	In Java, the package java.io is imported and the objects of class File, FileReader or FileWriter, are created. The objects of the classes access the inbuilt constructors and/ or methods of the classes through objects.	Python has several functions for creating, reading, updating, and deleting files. The key function for working with files in Python is the open() function. The open() function takes two parameters: filename and mode.
In C, the modes are r(read), w(write), and a(append), and these can be used directly.	In C++, the modes are in, out, and bin, and these are used with the help of scope resolution operators like ios::in and ios::out.	Java uses the concepts of streams, which are sequences of data.	In Python, the modes are r(read), w(write), a(append), and x(create). In addition, we can specify the file to be in b(binary) or t(text) mode.

11.6 FILE ORGANIZATION

File organization refers to the way in which records are physically arranged on a storage device. Further, there may be a single key or multiple keys associated with it. Therefore, based on its physical storage and the keys used to access the records, files are classified as sequential files, relative files, indexed sequential files, and inverted files. There are various factors which should be taken into consideration while choosing a particular type of file organization:

1. Ease of retrieval of the records

2. Economy of storage

3. Reliability, that is, whether a file organization is reliable

4. Security, that is, whether a file organization is secured

11.7 SEQUENCE FILE ORGANIZATION

Sequence file organization is the most basic way to organize a collection of records in a file. Sequence file organization is when the file is created when the records are written, one after the other in order, and can be accessed only in that order in which they are written when the file is used for input. All the records are numbered from zero onward. Thus, if there are N records in a file, then the first record is numbered 0, and the last record is numbered N–1. In some cases, records of sequential files are sorted by the value of some field in each record. The field whose value is used to sort the records is known as a *sort key*. If a file is sorted by the value of a field named "key field," then the record i proceeds record j, if, and only if, the value of "key field" in record i is less than or equal to the value of "key field" in record j. A file can be sorted in either ascending or descending order by a sort key comprising one or more fields. As the records in a sequential file can only be accessed sequentially, these files are used more commonly in batch processing than in interactive processing. For example, the records of a sequential file are used to generate the white pages of a telephone directory that are sorted by the subscriber's last name.

Advantages of a Sequence File Organization

1. It is easy to handle.

2. It does not involve extra overhead/problems.

3. Records can be of varying lengths in this organization.

4. It can be stored on magnetic disks as well as tapes.

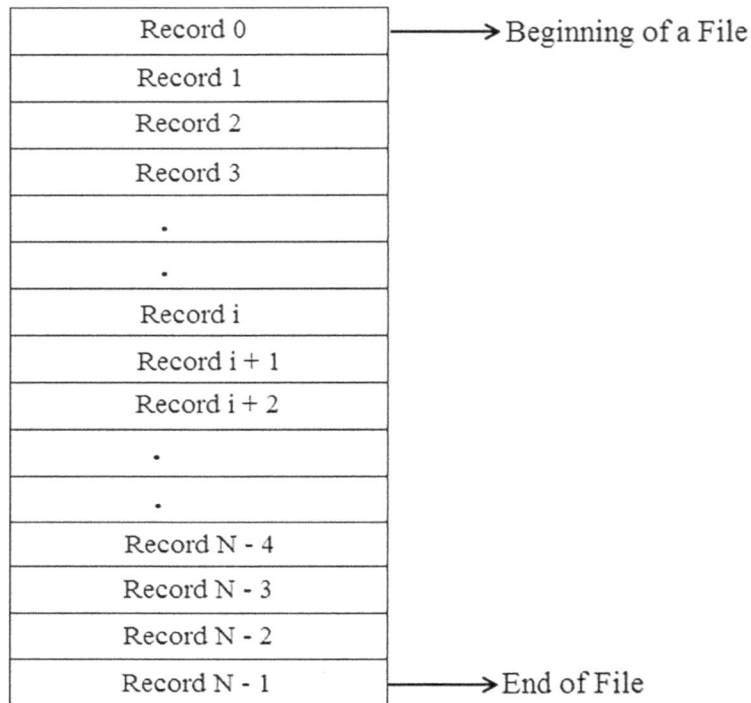

FIGURE 11.2 Structure of a sequence file organization

Disadvantages of Sequence File Organization

1. Records can be accessed only in sequence.

2. It does not support the update operation between files.

3. It does not support interactive applications.

11.8 INDEXED SEQUENCE FILE ORGANIZATION

An indexed sequential file organization is an efficient way of organizing the records when there is a need to access both sequentially by some key values and to access the records individually by the same key value. It provides the combination of access types that are supported by a sequential file or a relative file. The index is structured as a binary search tree. This index is used to serve as a request for access to a particular record, and the sequential data file alone is used to support sequential access to the entire collection of records. Because of its capability to support both sequential and direct access, indexed sequence file organization is used to support applications that require both batch and interactive processing.

Advantages of Indexed Sequence File Organization

1. Records can be accessed sequentially and randomly.

2. It supports batch and interactive oriented applications.

3. It supports the update operation between records in a file.

Disadvantages of Indexed Sequence File Organization

1. In this organization, files can only be stored on magnetic disks.

2. It involves extra overhead in the form of maintenance.

3. Records can only be of a fixed length, as we maintain the structure of each node like a linked list.

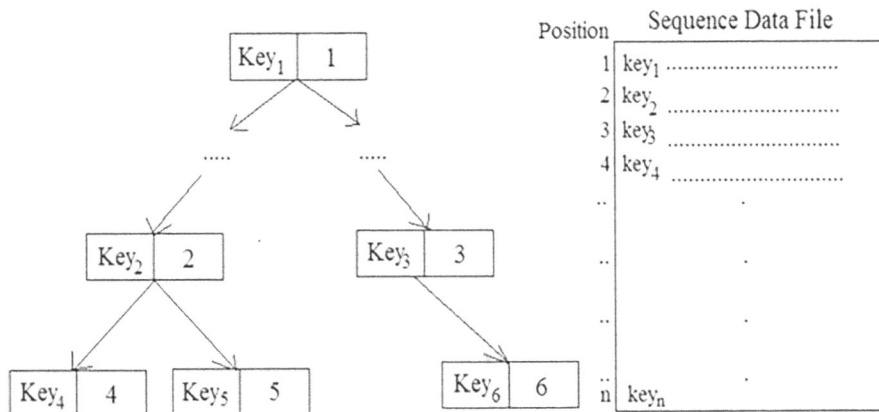

FIGURE 11.3 Use of BST and sequential files to provide indexed sequential access

11.9 RELATIVE FILE ORGANIZATION

Relative file organization provides an effective way of accessing individual records directly. In relative file organization, there is a predictable relationship between the key and the record's location in the file. The records do not necessarily appear physically in sorted order by their keys. Then how is a given record found? The relationship that is used to translate between the key value and the physical address is designated, for example, R(Key value → address). When a record is to be written into a relative file, the mapping function R is used to translate the record's key to an address, which indicates where the record is to be stored. The fundamental techniques that are used for mapping function R are the directory lookup and address calculation (hashing).

- **Directory Lookup Technique** – This is the simplest technique for implementing a mapping function R. The basic idea of this technique is to keep a directory of key values: address pairs. To find a record in a relative file, one locates its key value in the directory, and then the indicated address is used to find the record on the storage device. The directory can be organized as a binary search tree.

- **Address Calculation Technique** – Another common technique for implementing a mapping function R is to perform a calculation on the key value (hashing) such that the result is a relative address.

Advantages of Relative File Organization

1. Records can be accessed out of sequence.

2. It is well suited for interactive applications.

3. It supports an update operation in between the files.

Disadvantages of Relative File Organization

1. It can be stored only on magnetic disks.

2. It also involves extra overhead in the form of the maintenance of the indexes.

11.10 INVERTED FILE ORGANIZATION

One fundamental approach for providing a linkage between an index and a file is called inversion. A key's inversion index contains all the values that the key presently has in the records of the file. Each key-value entry in the inversion index points to all the data records that have the corresponding value. Then, the file is said to be *inverted* on that key. The inversion approach for providing multi-key access has been used as the basis for a physical data structure in commercially available relational DBMSs such as Oracle and DB2. These systems were designed to provide rapid access to the records via as many inversion keys as the designer cared to identify. They have user-friendly, natural-language-like query languages to assist the user in formulating inquiries. A complete inverted file has an inversion index for every data field. If a file is not completely inverted but has at least one inversion index, then it is said to be a *partially inverted file*.

Advantages of Inverted File Organization

1. The Boolean query requires only one access per record, satisfying the query, along with some access to process the indexes.

2. Records can be stored in any way, for example, sequentially ordered by primary key, randomly linked ordered by primary key, and so forth.

3. It also saves space as compared to other file structures.

Disadvantages of Inverted File Organization

Since the index entries are of variable lengths, index maintenance is rather complex.

11.11 SUMMARY

- A file is a collection of records. It is usually stored on a secondary storage device.

- The data is organized in a hierarchical order in the files. The hierarchical order includes items such as records and fields.

- File creation is the first operation to be performed on the files if the file is not created. A file is created by specifying its name and mode.

- A file is classified into two types, which are text files and binary files.

- A text file, often called a flat file, is a file that stores all the numeric or non-numeric data using its corresponding ASCII values. The data can be a string of letters, numbers, or special symbols.

- A binary file is a file that contains all the data in the binary form of 1s and 0s, It stores the data in the same form as that of primary memory.

- A file organization refers to the way in which records are physically arranged on a storage device.

- Sequence file organization is the most basic way to organize a collection of records in a file. In sequence file organization, the file is created when the records are written, one after the other in order, and can be accessed only in that order in which they are written when the file is used for input. All the records are numbered from zero onward.

- An indexed sequential file organization is an effective way of organizing the records when there is a need to access both sequentially by some key values and individually by the same key value. It provides the combination of access types that are supported by a sequential file or a relative file.

- Relative file organization provides an effective way of accessing individual records directly. In a relative file organization, there is a predictable relationship between the key and the record's location in the file.

- One fundamental approach for providing a link between an index and a file is called inversion. The inversion approach for providing multi-key access has been used as the basis for the physical data structure.

11.12 EXERCISES

11.12.1 Review Questions

Q1. What is a file?

Q2. Why is there a need to store data in files? Explain.

Q3. What do you understand about the terms *record* and *field*?

Q4. Discuss various operations that can be performed on files.

Q5. Differentiate between a text file and a binary file.

Q6. Write a short note on file attributes.

Q7. What do you understand about file organization?

Q8. Explain sequential file organization.

Q9. What are inverted files?

Q10. Explain indexed sequential file organization.

Q11. Give the merits and drawbacks of indexed sequential file organization.

Q12. What is relative file organization? Discuss the advantages and disadvantages of relative file organization.

11.12.2 Multiple Choice Questions

Q1. A collection of related fields is called

    a. Data

    b. Record

    c. Field

    d. File

Q2. A file marked as _____ can't be modified or deleted.

    a. Hidden

    b. Read-only

    c. Archive

    d. None of these

Q3. Which of the following is often known as a flat file?

    a. Binary File

    b. Text File

    c. String File

    d. None of these

Q4. _____ is a collection of data organized in a fashion that facilitates various operations such as updating and retrieving.

    a. Record

    b. Data word

    c. Field

    d. File

Q5. Relative files be used both for random and sequential access.

    a. True

    b. False

    c. Not possible to comment

Q6. A file marked as _____ is not displayed in the directory.

    a. Read-only

    b. Archive

    c. Hidden

    d. None of these

Q7. A data field is characterized by

    a. Type

    b. Size

    c. Mode

    d. Both (a) and (b)

Q8. _____ is used to store a collection of files.

    a. Record

    b. Dictionary

    c. Directory

    d. System

12

GRAPHS

12.1 INTRODUCTION

We have studied the types of linear data structures that are widely used in various applications. However, the only non-linear data structure we have studied thus far is trees. In trees, we discussed the parent-child relationship in which one parent can have many children. But in graphs, this parent-child relationship is less restricted, that is, any complex relationship can exist. Thus, a tree can be generalized as a special type of graph. Therefore, a graph is a non-linear data structure that has a wide range of real-life applications. A *graph* is a collection of some vertices (nodes) and edges that connect these vertices. Figure 12.1 represents a graph.

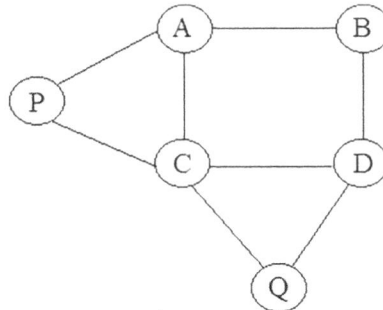

FIGURE 12.1 A graph

Thus, a graph G can be defined as an ordered set of vertices and edges (V, E), where V(G) represents the set of vertices, and E(G) represents the set of edges that connect these vertices. In the previous figure, V(G) = {A, B, C, D, P, Q} represents the set of vertices, and E(G) = {(A, B), (B, D), (D, C), (C, A), (C, Q), (Q, D), (A, P), (P, C)} represents the set of edges.

Practical Application:

A simple illustration of a graph is that when we connect with our friends on social media, where each user is a vertex and two users connect, forming an edge.

There are two types of graphs:

1. **Undirected Graph** – In an undirected graph, the edges do not have any direction associated with them. As we can see in the following figure, the two nodes A and B can be traversed in both the directions, that is, from A to B or from B to A. Thus, an undirected graph does not give any information about the direction.

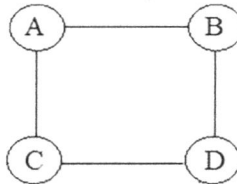

FIGURE 12.2 An undirected graph

2. **Directed Graph** – In a directed graph, the edges have directions associated with them. As we can see in the following figure, the two nodes A and B can be traversed in only one direction, that is, only from A to B and not from B to A. Therefore, in the edge (A, B), the node A is known as the *initial node* and node B is known as the *final node*.

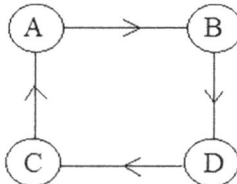

FIGURE 12.3 A directed graph

12.2 DEFINITIONS

- **Degree of a vertex/node** – The degree of a node is the total number of edges incident to that particular node. Here, the degree of node B is three, as three edges are incident to the node B.

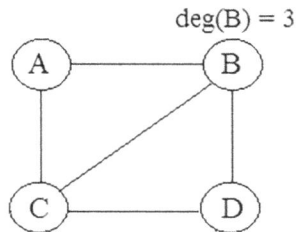

FIGURE 12.4 Graph showing the degree of node B

- **In-degree of a node** – The in-degree of a node is equal to the number of edges arriving at that particular node.

- **Out-degree of a node** – The out-degree is equal to the number of edges leaving that particular node.

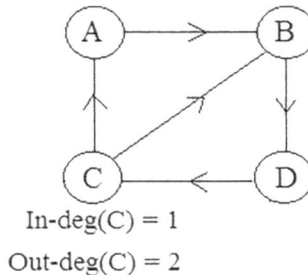

FIGURE 12.5 Graph showing in-degree and out-degree of node C

- **Isolated Node/Vertex** – A node having zero edges is known as the isolated node. The degree of such a node is zero.

FIGURE 12.6 Two isolated nodes, X and Y

- **Pendant Node/Vertex** – A node having one edge is known as a *pendant node*. The degree of such a node is one.

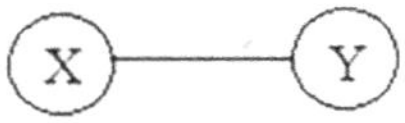

FIGURE 12.7 Two pendant nodes, X and Y

- **Adjacent Nodes** – For every edge e = (A, B) that connect nodes A and B, the nodes A and B are said to be the *adjacent nodes*.
- **Parallel Edges** – If there is more than one edge between the same pair of nodes, then they are known as *parallel edges*.

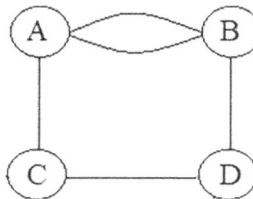

FIGURE 12.8 Parallel edges between A and B

- **Loop** – If an edge has a starting and ending point at the same node, that is, e = (A, A), then it is known as a *loop*.

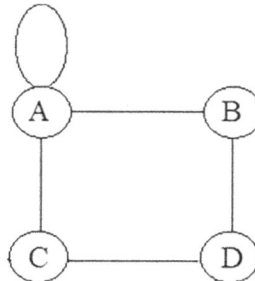

FIGURE 12.9 A loop

- **Simple Graph** – A graph G(V, E) is known as a simple graph if it does not contain any loops or parallel edges.

- **Complete Graph** – A graph G(V, E) is known as a complete graph if, and only if, every node in the graph is connected to another node and there is no loop on any of the nodes.

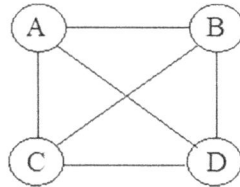

FIGURE 12.10 Complete graph

- **Regular Graph** – A regular graph is a graph in which every node has the same degree. If every node has a degree r, then the graph is called a regular graph of degree r. In the given figure, all the nodes have the same degree, that is, 2; hence, it is known as a 2-regular graph.

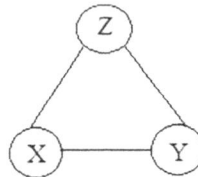

FIGURE 12.11 A 2-regular graph

- **Multi-graph** – A graph G(V, E) is known as a multi-graph if it contains either a loop, parallel edges, or both.

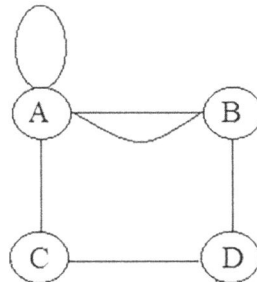

FIGURE 12.12 Multi-graph

- **Cycle** – This is a path containing one or more edges that start from a particular node and also terminate at the same node.
- **Cyclic Graph** – A graph that has cycles in it is known as a *cyclic graph*.

- **Acyclic Graph** – A graph without any cycles is known as an *acyclic graph*.
- **Connected Graph** – A graph G(V, E) is known as a *connected graph* if there is a path from any node in the graph to another node in the graph such that for every pair of distinct nodes, there must be a path.

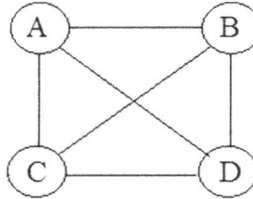

FIGURE 12.13 Connected graph

- **Strongly Connected Graph** – A directed graph is said to be a *strongly connected graph* if there exists a dedicated path between every pair of nodes in the graph. For example, if there are two nodes, say P and Q, and there is a dedicated path from P to Q, then there must be a path from Q to P.

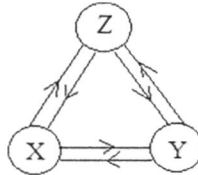

FIGURE 12.14 Strongly connected graph

- **Size of a Graph** – The size of a graph is equal to the total number of edges present in the graph.
- **Weighted Graph** – A graph G(V, E) is said to be a weighted graph if all the edges in the graph are assigned some data. This data indicates the cost of traversing the edge.

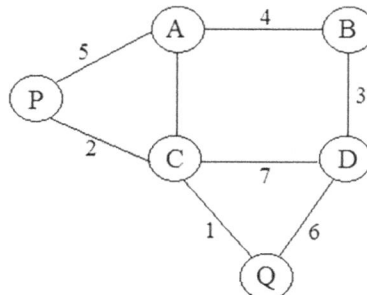

FIGURE 12.15 Weighted graph

12.3 GRAPH REPRESENTATION

Graphs can be represented in a computer's memory in either of the following ways:

1. Sequential Representation of Graphs using Adjacency Matrix

2. Linked Representation of Graphs using Adjacency List

12.3.1 Adjacency Matrix Representation

An adjacency matrix is used to represent the information of the nodes that are adjacent to one another. The two nodes are only adjacent when there is an edge connecting those nodes. For any graph G having n nodes, the dimension of the adjacency matrix is (n × n). Let G(V, E) be a graph having vertices V = {V_1, V_2, V_3.........V_n}, and then the adjacency matrix representation (n × n) is given by

$$a_{ij} = \begin{cases} 1 & \text{if there is an edge from } V_i \text{ to } V_j. \\ 0 & \text{otherwise} \end{cases}$$

The adjacency matrix is also known as a bit matrix or Boolean matrix since it contains only 0s and 1s. Now, let us take a few examples to discuss and understand it more clearly.

Example 1 – Consider the given directed graph and find its adjacency matrix.

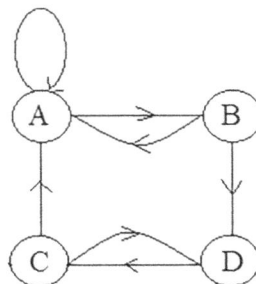

FIGURE 12.16 A directed graph

The adjacency matrix for the graph is

$$
\begin{array}{c c}
& \begin{array}{c c c c} A & B & C & D \end{array} \\
\begin{array}{c} A \\ B \\ C \\ D \end{array} &
\left[\begin{array}{c c c c}
1 & 1 & 0 & 0 \\
1 & 0 & 0 & 1 \\
1 & 0 & 0 & 1 \\
0 & 0 & 1 & 0
\end{array}\right]
\end{array}
$$

Example 2 – Now, consider the given undirected graph and find its adjacency matrix.

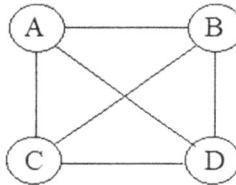

FIGURE 12.17 An undirected graph

The adjacency matrix for the graph is

$$
\begin{array}{c c}
& \begin{array}{c c c c} A & B & C & D \end{array} \\
\begin{array}{c} A \\ B \\ C \\ D \end{array} &
\left[\begin{array}{c c c c}
0 & 1 & 1 & 1 \\
1 & 0 & 1 & 1 \\
1 & 1 & 0 & 1 \\
1 & 1 & 1 & 0
\end{array}\right]
\end{array}
$$

Example 3 – Now, consider the given weighted graph and find its adjacency matrix.

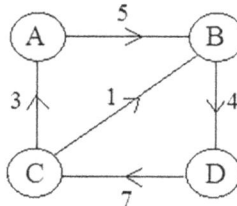

FIGURE 12.18 A directed weighted graph

The adjacency matrix for the graph is

$$
\begin{array}{c}
\\
A \\
B \\
C \\
D
\end{array}
\begin{bmatrix}
\begin{array}{cccc}
A & B & C & D \\
0 & 5 & 0 & 0 \\
0 & 0 & 0 & 4 \\
3 & 1 & 0 & 0 \\
0 & 0 & 7 & 0
\end{array}
\end{bmatrix}
$$

Example 4 – Consider the given undirected multi-graph and find its adjacency matrix.

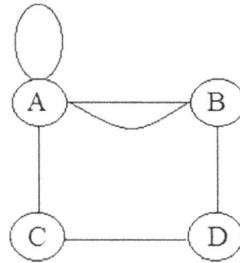

FIGURE 12.19 An undirected multi-graph

The adjacency matrix for the graph is

$$
\begin{array}{c}
\\
A \\
B \\
C \\
D
\end{array}
\begin{bmatrix}
\begin{array}{cccc}
A & B & C & D \\
1 & 2 & 1 & 0 \\
2 & 0 & 0 & 1 \\
1 & 0 & 0 & 1 \\
0 & 1 & 1 & 0
\end{array}
\end{bmatrix}
$$

From the previous examples, we conclude that

- The memory space needed to represent a graph using its adjacency matrix is n^2 bits.
- The adjacency matrix for an undirected graph is always symmetric.
- The adjacency matrix for a directed graph needs not to be symmetric.

- The adjacency matrix for a simple graph having no loops or parallel edges always contains 0s on the diagonal.
- The adjacency matrix for a weighted graph always contains the weights of the edges connecting the nodes instead of 0 and 1.
- The adjacency matrix for an undirected multi-graph contains the number of edges connecting the vertices instead of 1.

12.3.2 Adjacency List Representation

The adjacency matrix representation has some major drawbacks. First, it is very difficult to insert and delete the nodes in/from the graph as the size of the matrix needs to be changed accordingly, which is a very time-consuming process. Sometimes the matrix may contain many zeroes (sparse matrix). Hence, it is not a healthy representation. Therefore, adjacency list representation is preferred for representing sparse graphs in the memory. In this representation, every node is linked to its list of all the other nodes that are adjacent to it. Adjacency list representation makes it easier to add or delete nodes. It shows the adjacent nodes of a particular node. Now, let us consider a few examples.

Example 1 – Consider the given undirected graph and find its adjacency list representation.

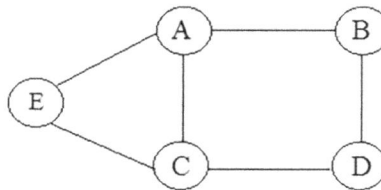

FIGURE 12.20 An undirected graph

The adjacency list representation of the graph is

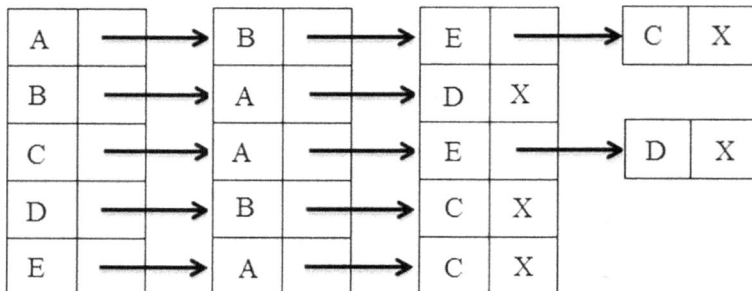

Example 2 – Consider the given directed graph and find its adjacency list representation.

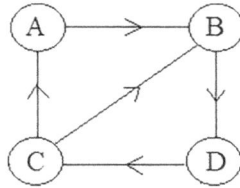

FIGURE 12.21 A directed graph

The adjacency list representation of the graph is

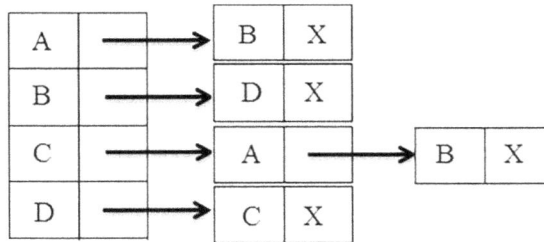

Example 3 – Now, consider the given weighted graph and find its adjacency list representation.

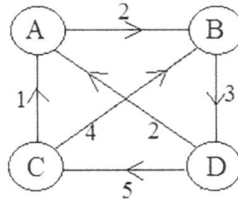

FIGURE 12.22 A directed weighted graph

The adjacency list representation of the graph is

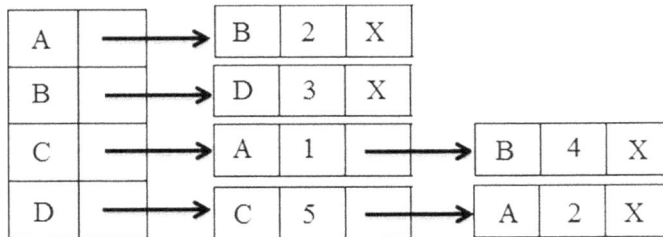

12.4 GRAPH TRAVERSAL TECHNIQUES

In this section, we discuss various types of techniques to traverse a graph. A graph is a collection of nodes and edges. Thus, traversing in a graph is the process of visiting each node and edge in some systematic approach. Therefore, there are two types of standard graph traversal techniques:

1. Breadth-First Search (BFS)

2. Depth-First Search (DFS)

12.4.1 Breadth-First Search

Breadth-First Search is a traversal technique that uses the queue as an auxiliary data structure for traversing all member nodes of a graph. In this technique, we select any node in the graph as a starting node, and then we take all the nodes adjacent to the starting node. We maintain the same approach for all the other nodes. We maintain the status of all the traversed/visited nodes in a queue so that no nodes are traversed again. Now, let us take a graph and apply BFS to traverse the graph.

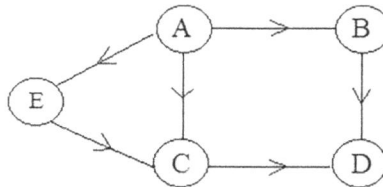

FIGURE 12.23 A sample graph

We start the traversal of the graph by taking node A as a starting node of the previous sample graph. Then, we traverse all the nodes adjacent to the starting node A. As we can see, B, C, and E are the adjacent nodes of A. So, we traverse these nodes in any order, say E, C, B. The traversal is

$$\boxed{A, E, C, B}$$

Now, we traverse all the nodes adjacent to E. Node C is adjacent to node E. But node C has already been traversed, so we ignore it and we move to the next step. We traverse all the nodes adjacent to node C. As we can see, D is the adjacent node of C. So we traverse node D and the traversal is

$$\boxed{A, E, C, B, D}$$

We can see that all the nodes have been traversed, and hence this was the Breadth-First Search traversal by taking node A as a starting node.

We implement the Breadth-First Search traversal technique with the help of a queue. In this, we maintain an array that stores all the adjacent unvisited neighbor nodes of a given node under consideration. Initially, the front and rear are set to –1. We also maintain the status of the visited nodes in a Boolean array, which has the value 1 if the node is visited and 0 if it is not visited.

- First, we en-queue/insert the starting node into the queue.
- Second, the first node/element in the queue is deleted from the queue, and all the adjacent unvisited nodes are inserted into the queue. This is repeated until the queue becomes empty.

For Example – Consider the following sample graph and traverse the graph using the breadth-first search technique.

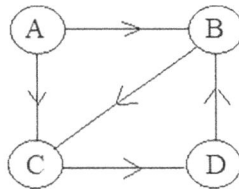

FIGURE 12.24 A sample graph

The appropriate adjacency list representation of the graph is given as follows:

Node	Adjacency List
A	B, C
B	C
C	D
D	B

In this example, we are taking A as a starting node.

Step 1: First, node A is inserted into the queue.

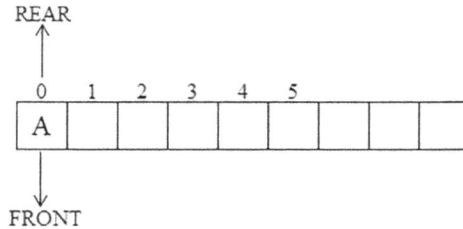

REAR

0	1	2	3	4	5		
A							

FRONT

Step 2: Node A is deleted from the queue and FRONT is incremented by 1. Now, insert all the nodes adjacent to A, which are nodes B and C, by incrementing REAR. Node A has also been traversed.

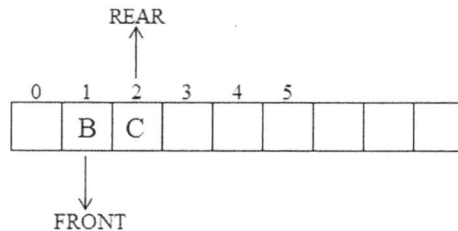

REAR

0	1	2	3	4	5		
	B	C					

FRONT

Step 3: Similarly, node B is deleted from the queue, and the FRONT is incremented by 1. Now, insert all the nodes adjacent to B, which is node C, by incrementing REAR. But C has already been inserted in the queue. So now, in this case, node C is also deleted by incrementing FRONT by 1, and the node adjacent to C, that is, D, is inserted into the queue. Therefore, nodes A, B, and C are traversed.

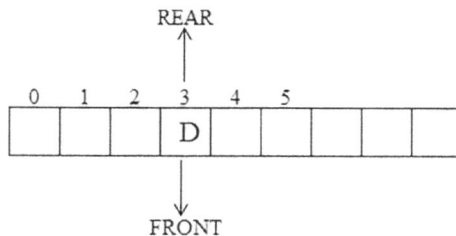

REAR

0	1	2	3	4	5		
			D				

FRONT

Step 4: Now, we again delete the front element from the queue which is D. We insert the adjacent node of D, that is, B. But it is already traversed.

Finally, as we delete the front element D, we notice that FRONT > REAR, which is not possible. Hence, we have traversed all the nodes in the graph.

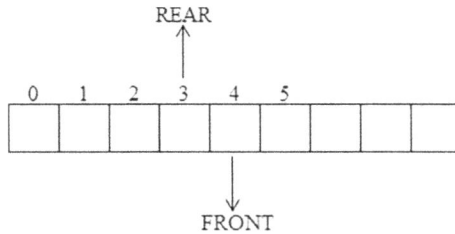

Therefore, the Breadth-First Search traversal of the graph is given as

A, B, C, D

Now, let us look at the function for a breadth-first search traversal. Here is a program for Breadth-First Search traversal

```python
# Python3 program to print a BFS traversal
# from a given source vertex. BFS(int s)
# traverses vertices reachable from s.
from collections import defaultdict

# This class represents a directed graph
# using adjacency list representation
class Graph:

    # Constructor
    def __init__(self):

        # default dictionary to store graph
        self.graph = defaultdict(list)

    # function to add an edge to graph
    def addEdge(self,u,v):
        self.graph[u].append(v)

    # Function to print a BFS of graph
    def BFS(self, s):

        # Mark all the vertices as not visited
        visited = [False] * (len(self.graph))

        # Create a queue for BFS
        queue = []
```

```
# Mark the source node as
# visited and enqueue it
queue.append(s)
visited[s] = True

while queue:

    # Dequeue a vertex from
    # queue and print it
    s = queue.pop(0)
    print (s, end = " ")

    # Get all adjacent vertices of the
    # dequeued vertex s. If a adjacent
    # has not been visited, then mark it
    # visited and enqueue it
    for i in self.graph[s]:
        if visited[i] == False:
            queue.append(i)
            visited[i] = True
```

The output of the program is as follows:

```
>>> g=Graph()
>>> g.addEdge(0,1)
>>> g.addEdge(0,2)
>>> g.addEdge(1,2)
>>> g.addEdge(2,0)
>>> g.addEdge(2,3)
>>> g.addEdge(3,3)
>>> g.BFS(2)
2 0 3 1
>>>
```

12.4.2 Depth-First Search

The Depth-First Search is another traversal technique that uses the stack as an auxiliary data structure for traversing all the member nodes of a graph. In this technique, we first select any node in the graph as a starting node, and then we travel along a path that begins from the starting node. We visit the adjacent node of the starting node, and again the adjacent node of the previous node, and so on. We maintain the same approach for all the other nodes. Now, let us take a graph and apply DFS to traverse the graph.

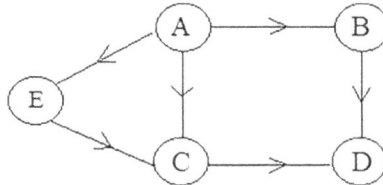

FIGURE 12.25 A sample graph

We start the traversal of the graph by taking node A as a starting node. Then, we traverse any of the nodes adjacent to the starting node A. As we can see, B, C, and E are the adjacent nodes of A. If we traverse node E, then we traverse the node adjacent to E, that is, C. After traversing C, we traverse the node adjacent to C, which is D. Now, there is no adjacent node to D; hence, we have reached the dead end. Thus, the traversal until now is

$$A, E, C, D$$

Because of the dead-end, we move backward. Now, we reach node C. We check if there is any other node adjacent to C. There is no such node, and thus we again move backward. Now, we reach E. We again check if there is any other node adjacent to E. There is no such node, and thus we again move backward. Now, we reach A. We check if there is any other node adjacent to A. There are two nodes, B and C, adjacent to node A. As C is already traversed, it will be ignored. Now, we traverse node B. After traversing B, we traverse the node adjacent to B, which is D, but D is already traversed. We can't move backward or forward. Thus, we have completed the traversal. The final traversal is given as

$$A, E, C, D, B$$

Now, we implement the Depth-First Search traversal technique with the help of a stack. In this, we maintain an array that stores all the adjacent unvisited neighbor nodes of a given node. Initially, the top is set to −1. We also maintain the status of the visited nodes in a Boolean array, which has the value of 1 if the node is visited and 0 if it is not visited.

- First, we push the starting node onto the stack.
- Second, the topmost node/element is popped out from the stack and is traversed. If it is already traversed, then we ignore it.
- Third, all the adjacent unvisited nodes of the popped node/element are pushed onto the stack. This process is repeated until the queue becomes empty. The steps are repeated until the stack becomes empty.

For Example – Consider the following sample graph and traverse the graph using the Breadth-First Search technique.

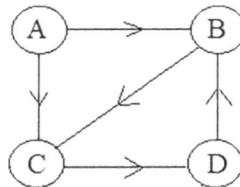

FIGURE 12.26 A sample graph

In this example, we take A as a starting node.

Step 1: Push A onto the stack.

A

Step 2: Now, pop the topmost element from the stack, that is, A. Thus, A is traversed. Now, push all the nodes adjacent to A, that is, push B and C.

B, C

Step 3: Again, pop the topmost element from the stack, that is, C. Thus, C is also traversed. Now, push all the nodes adjacent to C, that is, push D.

B, D

Step 4: Now, again pop the topmost element from the stack, that is, D. Thus, D is also traversed. Now, push all the nodes adjacent to D, that is, push B. But B is already in the stack. Therefore, no push is performed. Thus, the stack becomes

B

Step 5: Again, pop the topmost element from the stack, that is, B. Thus, B is also traversed. Now, push all the nodes adjacent to B, that is, push C. But C is already traversed; hence, the stack becomes empty.

Therefore, the depth-first search traversal of the graph is given as follows:

A, C, D, B

Now, let us look at the function for the depth-first search traversal.

Here is a program for Depth-First Search traversal.

```
# Python3 program to print DFS traversal
# from a given graph
from collections import defaultdict

# This class represents a directed graph using
# adjacency list representation
class Graph:

    # Constructor
    def __init__(self):

        # default dictionary to store graph
        self.graph = defaultdict(list)

    # function to add an edge to graph
    def addEdge(self, u, v):
        self.graph[u].append(v)

    # A function used by DFS
    def DFSUtil(self, v, visited):

        # Mark the current node as visited
        # and print it
        visited[v] = True
        print(v, end = ' ')

        # Recur for all the vertices
        # adjacent to this vertex
        for i in self.graph[v]:
            if visited[i] == False:
                self.DFSUtil(i, visited)
    # The function to do DFS traversal. It uses
    # recursive DFSUtil()
    def DFS(self, v):
```

```
# Mark all the vertices as not visited
visited = [False] * (max(self.graph)+1)

# Call the recursive helper function
# to print DFS traversal
self.DFSUtil(v, visited)
```

The output of the program is as follows:

```
>>> g=Graph()
>>> g.addEdge(0,1)
>>> g.addEdge(0,2)
>>> g.addEdge(1,2)
>>> g.addEdge(2,0)
>>> g.addEdge(2,3)
>>> g.addEdge(3,3)
>>> g.DFS(3)
```

```
>>> g.DFS(2)
2 0 1 3
>>> |
```

Memory Aid:

To remember which of the data structures are used in implementing a Breadth-First Search and Depth-First Search, we can use this memory aid. Breadth-First Search is implemented using a queue data structure, and a Depth-First Search is implemented using a stack data structure, as it can be remembered by alphabetical order. B (Breadth-First Search) and Q (Queue) come before than D (Depth-First Search) and S (Stack) in alphabetical order.

12.5 TOPOLOGICAL SORT

Topological sort is a procedure to determine the linear ordering of the nodes of an acyclic directed graph also known as (DAG) in which each node comes before all those nodes that have zero predecessors. A topological sort of a DAG is a linear ordering of the vertices of a graph G(V, E) such that if(a, b) is an edge, then a must appear before b in the topological ordering. The main idea behind this is that in a graph, if a vertex has in-degree 0, then that vertex should be selected as the first element in the topological order. A topological sort is possible only in acyclic directed graphs. An acyclic graph is one that does not have any cycles in it. Topological sorting is widely used in scheduling tasks, applications, and so on. Now, let us look at the algorithm for topological sorting.

Algorithm for Topological Sort

```
Step 1: START

Step 2: Find the in-degree of every node.

Step 3: Insert all the nodes/elements having in-degree zero
in the queue.

Step 4: Repeat Steps 5 and 6 until the queue becomes empty.

Step 5: Delete the first node from the queue by incrementing
FRONT by 1.

Step 6: Repeat for each neighbor P of node N -

        a) Delete the edge from P to M by decreasing the in-
           degree by 1.

        b) If the in-degree of P is zero, then add P to the
           rear of the queue.

Step 7: END
```

For Example – Consider a given acyclic directed graph and find its topological sort.

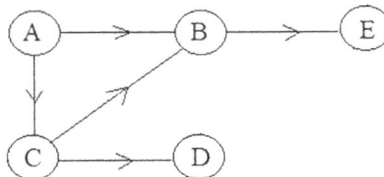

FIGURE 12.27 A cyclic directed graph

The appropriate adjacency list representation of the previous graph is given as follows:

Node	Adjacency List
A	B, C
B	E
C	B, D
D	-
E	-

Step 1: The in-degree of all the nodes is as follows:

In-degree (A) – 0
In-degree (B) – 2
In-degree (C) – 1
In-degree (D) – 1
In-degree (E) – 1

Now, we have node A with in-degree = 0; thus, A is added to the queue.

Step 2: Insert node A into the queue.

FRONT = 1, REAR = 1, QUEUE = A

Step 3: Delete node A from the queue. Delete all the edges going from A.

FRONT = 0, REAR = 0, TOPOLOGICAL SORT = A
Thus, the graph becomes

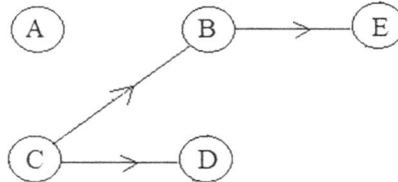

The in-degree of all the nodes is as follows:

In-degree (B) – 1
In-degree (C) – 0
In-degree (D) – 1
In-degree (E) – 1

Now, we have node C with in-degree = 0; thus, C is added to the queue.

Step 4: Insert node C into the queue.

FRONT = 1, REAR = 1, QUEUE = C

Step 5: Delete node C from the queue. Delete all the edges going from C.

FRONT = 0, REAR = 0, TOPOLOGICAL SORT = A, C
Thus, the graph becomes

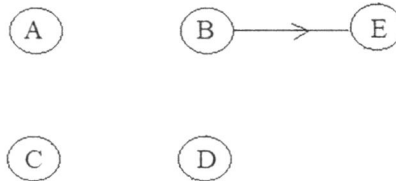

The in-degree of all the nodes is as follows:

In-degree (B) – 0
In-degree (D) – 0
In-degree (E) – 1

Now, we have two nodes B and D with in-degree = 0; thus, B and D are added to the queue.

Step 6: Insert nodes B and D into the queue.

FRONT = 1, REAR = 2, QUEUE = B, D

Step 7: Delete node B from the queue. Delete all the edges going from B. There will be no change in the in-degree of the nodes.

FRONT = 1, REAR = 1, TOPOLOGICAL SORT = A, C, B, QUEUE = D

Step 8: Delete node D from the queue. Delete all the edges going from D.

FRONT = 0, REAR = 0, TOPOLOGICAL SORT = A, C, B, D
Thus, the graph becomes

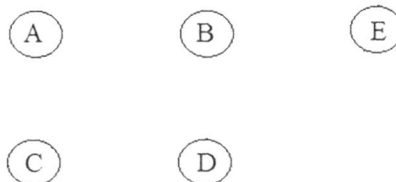

The in-degree of all the nodes is as follows:

In-degree (E) – 0

Now, we have node E with in-degree = 0. Thus, E is added to the queue.

Step 9: Insert node E into the queue.

FRONT = 1, REAR = 1, QUEUE = E

Step 10: Delete node E from the queue. Delete all the edges going from E.

FRONT = 0, REAR = 0, TOPOLOGICAL SORT = A, C, B, D, E

Now, we have no nodes left in the graph. Thus, the topological sort of the graph is

$$\boxed{\text{A, C, B, D, E}}$$

12.6 MINIMUM SPANNING TREE

A spanning tree of an undirected and connected graph G is a subgraph that contains all the vertices and edges that connect these vertices and is a tree. The weights/costs can be assigned to the edges, and these weights/costs can be used to calculate the weight/cost of the spanning tree by calculating the sum of the weights/costs of each edge. A graph can have many spanning trees. Thus, a Minimum Spanning Tree (MST) is defined as a spanning tree that has weights/costs associated with the edges such that the total weight/cost of the spanning tree is at a minimum. Although there are various approaches for determining an MST, the two most popular approaches for determining a minimum cost spanning tree of a graph are as follows:

1. Prim's Algorithm

2. Kruskal's Algorithm

12.6.1 Prim's Algorithm

Prim's algorithm is the algorithm that is used to build a minimum cost spanning tree. This algorithm works in such a way that it builds a tree edge by edge. The next edge to be included is chosen according to some criteria. The steps involved in Prim's algorithm are as follows:

Step 1: Select a starting vertex/node and add it to the spanning tree.

Step 2: During each iteration, select a vertex/node in such a way that the edge connecting vertex V_i to another vertex V_j has the minimum cost/weight assigned to it. Remember, the edge forming a cycle must not be added.

Step 3: End the process when $(n-1)$ number of edges have been inserted into the tree.

Frequently Asked Questions

Q. Consider the given graph and construct a minimum spanning tree using Prim's algorithm.

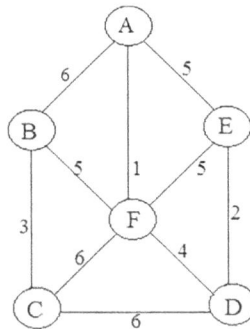

Answer:

Step 1: The starting node is F.

Step 2: The lowest weighted/cost edge is (F, A), that is, 1. Hence, it is added to the tree.

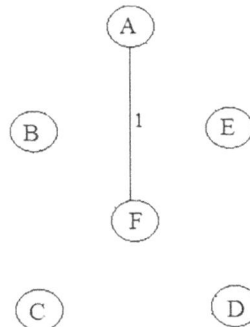

Step 3: *Now, the lowest weighted/cost edge is (F, D), that is, 4. Hence, it is added to the tree.*

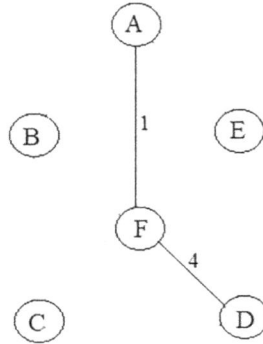

Step 4:

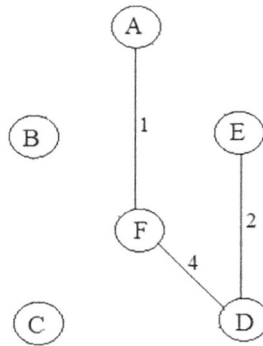

Step 5:

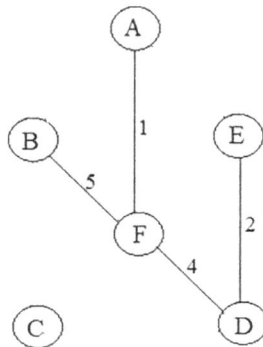

Step 6: *Finally, the minimum spanning tree is constructed.*

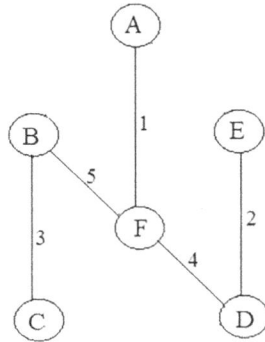

12.6.2 Kruskal's Algorithm

Kruskal's algorithm is another approach for determining the minimum cost spanning tree of a graph. In this approach also, the tree is built edge by edge. The next edge to be included is chosen according to some criteria. The steps involved in Kruskal's algorithm are as follows:

Step 1: The weights/costs assigned to the edges are sorted in ascending order.

Step 2: In this step, the lowest weighted/cost edge is added to the tree. Remember, the edge forming a cycle must not be added.

Step 3: End the process when $(n-1)$ number of edges have been inserted into the tree.

Frequently Asked Questions

Q. Consider the given graph and construct a minimum spanning tree using Kruskal's algorithm.

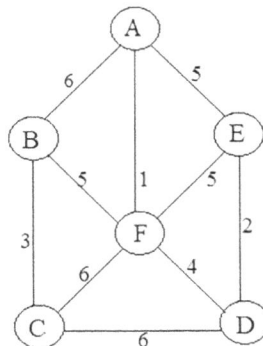

Answer:

Step 1: Initially the tree is given as

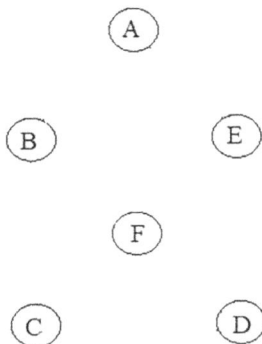

Step 2: Choose edge (F, A).

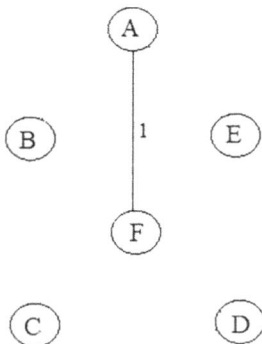

Step 3: Choose edge (D, E).

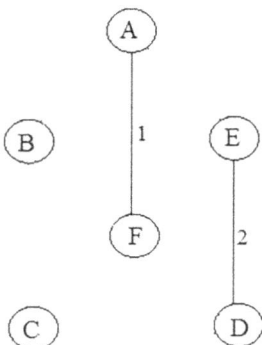

Step 4: *Choose edge (B, C).*

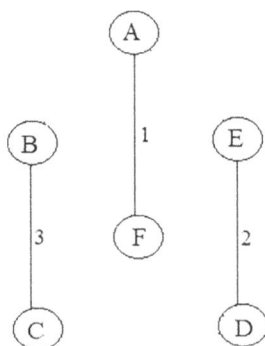

Step 5: *Choose edge (F, D).*

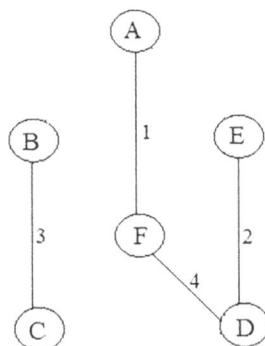

Step 6: *Choose edge (F, B).*

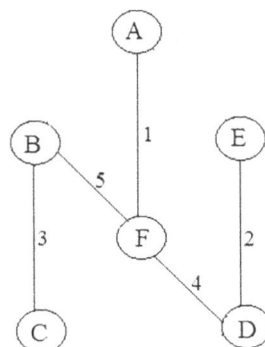

Practical Application:

Graphs are used to find the shortest route for GPS, Google maps, and Yahoo maps.

12.7 SUMMARY

- A graph is a collection of vertices (nodes) and edges that connect these vertices.

- The degree of a node is the total number of edges incident to that particular node.

- A graph G (V, E) is known as a complete graph if, and only if, every node in the graph is connected to another node and there is no loop on any of the nodes.

- An adjacency matrix is usually used to represent the information of the nodes which are adjacent to one another. The adjacency matrix is also known as a bit matrix or Boolean matrix because it contains only 0s and 1s.

- In adjacency list representation, every node is linked to its list of all the other nodes which are adjacent to it.

- Traversing in a graph is the process of visiting each node and edge in some systematic approach.

- Breadth-First Search is a traversal technique that uses the queue as an auxiliary data structure for traversing all the member nodes of the graph. In this technique, we select any node in the graph as a starting node, and then we take all the nodes adjacent to the starting node. We maintain the same approach for all the other nodes.

- The Depth-First Search is another traversal technique that uses the stack as an auxiliary data structure for traversing all the member nodes of the graph. In this also, we select any node in the graph as a starting node, and then we travel along a path that begins from the starting node. We visit the adjacent node of the starting node, and again the adjacent node of the previous node, and so on.

- Topological sort is a procedure to determine the linear ordering of the nodes of an acyclic directed graph also known as (DAG) in which each node comes before all those nodes which have zero predecessors.

- A Minimum Spanning Tree (MST) is defined as a spanning tree that has weights/costs associated with the edges such that the total weight/cost of the spanning tree is at a minimum.

12.8 EXERCISES

12.8.1 Theory Questions

Q1. What is a graph? Explain its features.

Q2. What do you understand about a complete graph?

Q3. What is a multi-graph?

Q4. How can a graph be represented in the computer's memory?

Q5. Differentiate between a directed and undirected graph with an example of each.

Q6. Consider the following graph and find the following:

    a. Adjacency Matrix Representation

    b. Degree of each node

    c. Is the graph complete?

    d. Pendant nodes

Q7. Explain why adjacency list representation is preferred for storing sparse matrices over adjacency matrix representation.

Q8. What are the different types of graph traversal techniques? Explain each of them in detail with the help of an example.

Q9. What do you understand about topological sort?

Q10. In what kind of graphs can topological sorting be used?

Q11. Differentiate between the Breadth-First Search and Depth-First Search.

Q12. Consider the following graph and find out its BFS and DFS traversal.

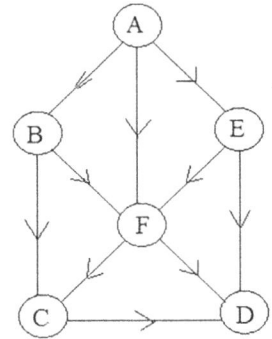

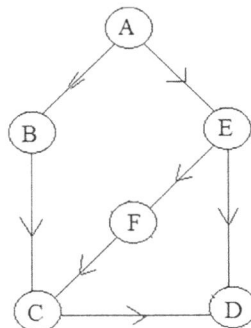

Q13. What is a spanning tree?

Q14. Why is a minimum spanning tree called a spanning tree? Discuss.

Q15. Consider the given adjacency matrix and draw the directed graph.

$$\begin{array}{c} & \begin{array}{cccc} A & B & C & D \end{array} \\ \begin{array}{c} A \\ B \\ C \\ D \end{array} & \left[\begin{array}{cccc} 1 & 1 & 0 & 1 \\ 1 & 1 & 1 & 0 \\ 0 & 1 & 1 & 0 \\ 0 & 1 & 1 & 1 \end{array} \right] \end{array}$$

Q16. Write a short note on Prim's algorithm.

Q17. Explain Kruskal's algorithm.

Q18. List some of the real-life applications of graphs.

Q19. Consider the following graph and find the minimum spanning tree using

    a. Prim's algorithm

    b. Kruskal's algorithm

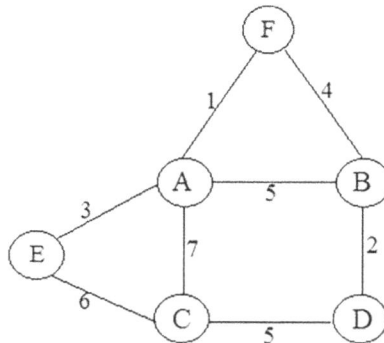

12.8.2 Programming Questions

Q1. Write a Python program to create and display a graph.

Q2. Write an algorithm to perform a topological sort on a graph.

Q3. Write an algorithm to find the degree of a node N in a graph.

Q4. Write a Python program to traverse a graph using a Depth-First Search.

Q5. Write an algorithm to traverse a graph using the Breadth-First Search.

Q6. Write a Python program to find the shortest path using Prim's algorithm.

Q7. Write a Python program to find the shortest path using Kruskal's algorithm.

12.8.3 Multiple Choice Questions

Q1. To implement the breadth-first search, the data structure used is

- **a.** Stack
- **b.** Queue
- **c.** Trees
- **d.** Linked List

Q2. A graph having multiple edges is known as a _____.

- **a.** Connected Graph
- **b.** Complete Graph
- **c.** Simple Graph
- **d.** Multi-graph

Q3. An edge having initial and endpoints at the same node is called

- **a.** Degree
- **b.** Cycle
- **c.** Loop
- **d.** Parallel Edge

Q4. An adjacency matrix is also known as a

- **a.** Bit Matrix
- **b.** Boolean Matrix
- **c.** Both of the above
- **d.** None of the above

Q5. To implement the depth-first search, the data structure used is

  a. Stack

  b. Queue

  c. Trees

  d. Linked List

Q6. Topological Sort is performed only on

  a. Cyclic Directed Graphs

  b. Acyclic Directed Graphs

  c. Both of the above

  d. None of the above

Q7. Which one of the following nodes has a zero degree?

  a. Simple node

  b. Isolated node

  c. Pendant node

  d. None of the above

Q8. _____ is the total number of nodes in a graph.

  a. Degree

  b. In-degree

  c. Out-degree

  d. Size

Q9. Graph G can have many spanning trees.

  a. True

  b. False

  c. Not possible to comment

ANSWERS TO MULTIPLE CHOICE QUESTIONS

Chapter 1

1. c
2. b
3. d
4. a
5. d
6. d
7. c
8. d
9. b
10. d
11. a
12. c
13. a
14. c
15. d

Chapter 2

1. a
2. b
3. c
4. a
5. a
6. b
7. a
8. d
9. d
10. c

Chapter 3

1. b
2. c
3. d
4. a
5. a
6. b
7. d
8. a

Chapter 4

1. b
2. d
3. b
4. b
5. d
6. a
7. c
8. c
9. b

Chapter 5

1. c
2. a
3. b
4. a
5. b
6. c
7. c
8. b
9. b
10. a

Chapter 6

1. a
2. d
3. c
4. a
5. d
6. b
7. c
8. b
9. c

Chapter 7

1. b
2. b
3. c
4. a
5. b
6. a
7. c

8. d
9. c

Chapter 8

1. d
2. c
3. c
4. d
5. c
6. c
7. c
8. b
9. a
10. a
11. b
12. c
13. b
14. c
15. d

Chapter 9

1. b
2. a
3. d
4. d
5. c
6. c
7. d
8. b
9. d

Chapter 10

1. c
2. d
3. d
4. b
5. a
6. a
7. a
8. b
9. b

Chapter 11

1. b
2. b
3. a
4. d
5. a
6. c
7. d
8. c

Chapter 12

1. b
2. d
3. c
4. c
5. a
6. b
7. b
8. d
9. b

INDEX

www.ingramcontent.com/pod-product-compliance
Lightning Source LLC
Chambersburg PA
CBHW080703220326
41598CB00033B/5296